"Tells a number of fascinating stories about ever-cheaper, ever-more-powerful computer chips and the tools and techniques . . . that will increasingly and dramatically affect nearly every area of our lives."

— *Philadelphia Inquirer*

"Baker does an excellent job of breaking down the concepts into simple, easy-to-digest terms . . . The book is a wake-up call."

—*Rocky Mountain News*

"Most of us are at least faintly uneasy about the ubiquity of data compiled about our lives. We should be. *The Numerati* shows us just how far this process has already gone, how much further it has to go—and how little we can do to avoid it."

—*Financial Times*

"A fascinating and fast read. Baker has a knack for describing statistical techniques in ways that everyone can understand, without formulas and without jargon, while illustrating them with real-world issues."

—*National Review*

"A strikingly well-argued and positive account of a wired, watched world in which private lives are no longer an option."

—*Daily Telegraph*

THE NUMERATI

THE
NUMERATI

————>‹—————

S T E P H E N B A K E R

MARINER BOOKS
HOUGHTON MIFFLIN HARCOURT
BOSTON ✳ NEW YORK

First Mariner Books edition 2009

Copyright © 2008 by Stephen Baker

For information about permission to reproduce
selections from this book, write to Permissions,
Houghton Mifflin Harcourt Publishing Company,
215 Park Avenue South, New York, New York 10003.

www.hmhbooks.com

Library of Congress Cataloging-in-Publication Data
Baker, Stephen, date.
The numerati / Stephen Baker.
p. cm.
Includes bibliographical references.
ISBN 978-0-618-78460-8
1. Mathematical models—Social aspects. 2. Human
behavior—Mathematical models. 3. Mathematical
statistics—Data processing. I. Title.
QA401.B35 2008   303.48'3—dc22

ISBN 978-0-547-24793-9 (pbk.)

Book design by Melissa Lotfy

Printed in the United States of America

DOC 10 9 8 7 6 5 4 3 2 1

For Jalaire

CONTENTS

THE NUMERATI

INTRODUCTION

IMAGINE YOU'RE IN a café, perhaps the noisy one I'm sitting in at this moment. A young woman at a table to your right is typing on her laptop. You turn your head and look at her screen. She surfs the Internet. You watch.

Hours pass. She reads an online newspaper. You notice that she reads three articles about China. She scouts movies for Friday night and watches the trailer for *Kung Fu Panda*. She clicks on an ad that promises to connect her to old high school classmates. You sit there taking notes. With each passing minute, you're learning more about her. Now imagine that you could watch 150 million people surfing at the same time. That's more or less what Dave Morgan does.

"What is it about romantic-movie lovers?" Morgan asks, as we sit in his New York office on a darkening summer afternoon. The advertising entrepreneur is flush with details about our ramblings online. He can trace the patterns of our migrations, as if we were swallows or humpback whales, while we move from site to site. Recently he's become intrigued by the people who click most often on an ad for car rentals. Among them, the largest group had paid a visit to online obituary listings. That makes sense, he says, over the patter of rain against the windows. "Someone dies, so you fly to the funeral and rent a car." But it's the second-largest group that has Morgan scratching his head. Romantic-movie lovers. For some reason Morgan can't fathom, loads of them seem drawn to a banner ad for Alamo Rent A Car.

Morgan, a cheery 43-year-old, wears his hair pushed to the side, as if when he was young his mother dipped a comb into water, drew it across, and the hair just stayed there. He grew up in Clearfield, a small town in western Pennsylvania a short drive from Punxsutawney. Every year on the second day of February, halfway between the winter solstice and the vernal equinox, a crowd in that town gathers around a large caged rodent still groggy from hibernation. They study the animal's response to its own shadow. According to ancient Celtic lore, that single bit of data tells them whether spring will come quickly or hold off until late March. Morgan has migrated as far as can be from such folk predictions. At his New York start-up, Tacoda, he hires statisticians to track our wanderings on the Web and figure out our next moves. Morgan was a pioneer in Internet advertising during the dot-com boom, starting up an agency called 24/7 Real Media. During the bust that followed he founded another company, Tacoda, and moved seamlessly into what he saw as the next big thing: helping advertisers pinpoint the most promising Web surfers for their message.

Tacoda's entire business gorges on data. The company has struck deals with thousands of online publications, from the *New York Times* to *BusinessWeek*. Their sites allow Tacoda to drop a bit of computer code called a cookie into our computers. This lets Tacoda trace our path from one site to the next. The company focuses on our behavior and doesn't bother finding out our names or other personal details. (That might provoke a backlash concerning privacy.) But Tacoda can still learn plenty. Let's say you visit the *Boston Globe* and read a column on the Toyota Prius. Then you look at the car section on AOL. Good chance you're in the market for wheels. So Tacoda hits you at some point in your Web wanderings with a car ad. Click on it, and Tacoda gets paid by the advertiser — and gleans one more detail about you in the process. The company harvests 20 billion of these behavioral clues every day.

Sometimes Morgan's team spots groups of Web surfers who appear to move in sync. The challenge then is to figure out what triggers their movements. Once this is clear, the advertisers can anticipate people's online journeys — and sprinkle their paths with just the right ads. This requires research. Take the curious connection between fans of romance movies and the Alamo Rent A Car ad. To come to grips with it, Morgan and his colleagues have to dig deeper into the data. Do car renters arrive in larger numbers from a certain type of romance movie, maybe ones that take place in an exotic locale? Do members of this group have other favorite sites in common? The answers lie in the strings of ones and zeros that our computers send forth. Maybe the statistics will show that the apparent link between movie fans and car renters was just a statistical quirk. Or perhaps Morgan's team will unearth a broader trend, a correlation between romance and travel, lust and wanderlust. That could lead to all kinds of advertising insights. In either case, Morgan can order up hundreds of tests. With each one he can glean a little bit more about us and target the ads with ever more precision. He's taking analysis that once ran through an advertiser's gut, and replacing it with science. We're his guinea pigs — or groundhogs — and we never stop working for him.

. . .

WHEN IT COMES to producing data, we're prolific. Those of us wielding cell phones, laptops, and credit cards fatten our digital dossiers every day, simply by living. Take me. As I write on this spring morning, Verizon, my cell phone company, can pin me down within several yards of this café in New Jersey. Visa can testify that I'm well caffeinated, probably to overcome the effects of the Portuguese wine I bought last night at 8:19. This was just in time for watching a college basketball game, which, as TiVo might know, I turned off after the first half. Security cameras capture time-stamped images of me near every bank and convenience store. And don't get me started on my Web wanderings. Those are already a matter of record for dozens of Internet publishers and advertisers around the world. Dave Morgan is just one in a large and curious crowd. Late in the past century, to come up with this level of reporting, the East German government had to enlist tens of thousands of its citizens as spies. Today we spy on ourselves and send electronic updates minute by minute.

This all started with computer chips. Until the 1980s, these bits of silicon, bristling with millions of microscopic transistors, were still a novelty. But they've grown cheaper and more powerful year by year, and now manufacturers throw them into virtually anything that can benefit from a dab of smarts. They power our cell phones, the controls in our cars, our digital cameras, and, of course, our computers. Every holiday season, the packages we open bring more chips into our lives. These chips can record every instruction they receive and every job they do. They're fastidious note takers. They record the minutiae of our lives. Taken alone, each bit of information is nearly meaningless. But put the bits together, and the patterns describe our tastes and symptoms, our routines at work, the paths we tread through the mall and the supermarket. And these streams of data circle the globe. Send a friend a smiley face from your cell phone. That bit of your behavior, that tiny gesture, is instantly rushing, with billions of others, through fiber-optic cables. It's soaring up to a satellite and back down again and checking in at a server farm in Singapore before you've put the phone back in your pocket. With so many bits flying around, the very air we breathe is teeming with motes of information.

If someone could gather and organize these far-flung electronic gestures, our lives would pop into focus. This would create an ever-changing, up-to-the-minute mosaic of human behavior. The prospect is enough to make marketers quiver with excitement. Once they have a bead on our data, they can decode our desires, our fears, and our needs. Then they can sell us precisely what we're hankering for.

But it sounds a lot simpler than it is. Sloshing oceans of data, from e-mails and porn downloads to sales receipts, create immense chaotic waves. In a single month, Yahoo alone gathers 110 billion pieces of data about its customers, according to a 2008 study by the research firm comScore. Each person visiting sites in Yahoo's network of advertisers leaves behind, on average, a trail of 2,520 clues. Piece together these details, you might think, and our portraits as shoppers, travelers, and workers would gell in an instant. Summoning such clarity, however, is a slog. When I visit Yahoo's head of research, Prabhakar Raghavan, he tells me that most of the data trove is digital garbage. He calls it "noise" and says that it can easily overwhelm Yahoo's computers. If one of Raghavan's scientists gives an imprecise computer command while trawling through Yahoo's data, he can send the company's servers whirring madly through the noise for days on end. But a timely tweak in these instructions can speed up the hunt by a factor of 30,000. That reduces a 24-hour process to about three seconds. His point is that people with the right smarts can summon meaning from the nearly bottomless sea of data. It's not easy, but they can find us there.

The only folks who can make sense of the data we create are crack mathematicians, computer scientists, and engineers. They know how to turn the bits of our lives into symbols. Why is this necessary? Imagine that you wanted to keep track of everything you ate for a year. If you're like I was in the fourth grade, you go to the stationery store and buy a fat stack of index cards. Then, at every meal you write the different foods on fresh cards. Meat loaf. Spinach. Tapioca pudding. Cheerios. After a few days, you have a growing pile of cards. The problem is, there's no way to count or analyze them. They're just a bunch of words. These are symbols too, of course, each one representing a thing or a concept. But they are near impossible to add or subtract, or to drop into a graph illustrating a trend. Put these words in a pile, and they add up to what the specialists call "unstructured data." That's computer talk for "a big mess." A better approach would be to label all the meats with *M,* all the green vegetables with *G,* all the candies with *C,* and so on. Once the words are reduced to symbols, you can put them on a spreadsheet and calculate, say, how many times you ate meat or candy in a given week. Then you can make a graph linking your diet to changes in your weight or the pimple count on your face.

The key to this process is to find similarities and patterns. We humans do this instinctively. It's how we figured out, long ago, which plants to eat and how to talk. But while many of us were focusing on specific challenges, others were thinking more symbolically. I picture early humans sitting around a fire. Some, naturally, are jousting for the biggest piece of meat or busy with mating rituals. But off to the side, a select few are toying with stones, thinking, "If each of these pebbles represents one mammoth, then this rock . . ." Later, notes Tobias Dantzig in *Number: The Language of Science,* the Romans used their word *calcula,* meaning "pebble," to give a name to this thought process. But the pebble was just the start. The essence of calculation was to advance from the physical pebbles to ever-higher realms of abstract reasoning.

That science developed over the centuries, and we now have experts who are comfortable working with ridiculously large numbers, the billions and trillions that the rest of us find either unimaginable or irrelevant. They are heirs to the science that turns our everyday realities into symbols. As the data we produce continues to explode and computers grow relentlessly stronger, these maestros gain in power. Two of them made a big splash in the late 1990s by founding Google. For the age we're entering, Google is the marquee company. It's built almost entirely upon math, and its very purpose is to help us hunt down data. Google's breakthrough, which transformed a simple search engine into a media giant, was the discovery that our queries—the words we type when we hunt for Web pages—are of immense value to advertisers. The company figured out how to turn our data into money. And lots of others are looking to do the same thing. Data whizzes are pouring into biology, medicine, advertising, sports, politics. They are adding us up. We are being quantified.

When this process began, a half-century ago, the first computers were primitive boxes the size of a garbage truck. They kept their distance from us, purring away in air-conditioned rooms. At this early stage, the complexity of the human animal was too much for them. They couldn't even beat us at chess. But in certain numerical domains, they showed promise. An early test involved consumer credit. In 1956, two Stanford graduates, a mathematician named Bill Fair and his engineer friend Earl Isaac, came up with the idea of replacing loan officers with a computer. This hulking machine knew practically nothing, not even what the applicants did for a living. It certainly hadn't learned if they'd gotten a raise or filed for divorce. Legions of human loan officers, by contrast, were swimming in data. They often knew the families of the loan applicants. They were acquainted with how much the applicant had struggled in high school and how his engagement had fallen through, probably because of a drinking problem (if he was anything like his uncle). The loan officers had enough details to write sociological monographs, if they were so inclined, about the families in their towns. But they lacked a scientific system to analyze it all. Bankers depended, for the most part, on their gut.

By contrast, the computerized approach zeroed in on only a small set of numbers, most of them concerning bank balances, debts, and payment history. Bare bones. Fair and Isaac built a company to analyze the patterns of those numbers. They developed a way to determine the odds that each customer would default on a loan. Everyone got a number. These risk scores proved to be much better predictors than the gut-trusting humans. Most borrowers with high credit scores made good on their loans. And more people qualified for them. The machine, after all, didn't discriminate on the basis of anything but numbers. It was equal-opportunity banking. Like a lot of analytical systems, it was fairer. Its narrow scope, paradoxically, returned broad-minded results. What's more, a lot of people turned out to be better bets than the loan officers suspected. The market for credit expanded.

Still, the computer knew its place. It thrived in the world

of numbers, and it stayed there. Those of us who specialized in words and music and images barely noticed it. Yet over the following decades, the computer grew in power, gobbling up ever more ones and zeros per millisecond. It got cheaper and smaller, and it linked up with others around the world. It produced jaw-dropping efficiencies. And from the viewpoint of the humanities crowd (including this history major), it swallowed entire technologies. It supplanted typewriters and moved on, like an imperial force, to rout record players and film cameras. It took over the mighty telephone. Finally, in the 1990s, even those of us who had long viewed computers as aliens from the basement world of geekdom started to make room for them in our homes and offices. We learned that we could use these machines to share our words and movies and photos with the entire world.

In fact, we had little choice. The old ways were laughably slow. But there was one condition: we had to render everything we sent, the very stuff of our lives, into ones and zeros. That's how we came to deliver our riches, the key to communications on earth, to the masters of the symbolic language. Now these mathematicians and computer scientists are in a position to rule the information of our lives. I call them the Numerati.

ON A SWELTERING summer afternoon, Dave Morgan sits in his Spartan office overlooking Seventh Avenue. He has the shade drawn to keep out the heat, and he can't figure out how to turn on the fluorescent light. Sitting in the shadows, he tells me how marketing has changed over the past generation. Traditionally, he says, marketers concentrated on big groups of us. We weren't much more diverse, from their point of view, than the lines of General Motors cars: Cadillacs and Buicks for the rich and wannabes, Chevys for the middle class, Pontiacs for young hotshots, and pickup trucks for farmers. They didn't need to know much more than that because midcentury American factories, whether they were producing blue jeans or peanut butter, were manufacturing mass quantities. Smaller, more focused runs cost too much money. Sure, certain neighborhoods in coastal cities attracted eccentrics who drove foreign cars and walked around in lederhosen or berets. But for the most part, we ate, wore, and drove what the mass-production factories churned out, and we learned about it through the mass media. This model, created in the United States, spread in the decades following World War II across Europe and to much of Asia and Latin America. It was an efficient way to reach millions of consumers with machine-made goods.

Advertising in this industrial complex, Morgan says, was simple. You cut the population into five or six demographic groups, based on income, gender, and neighborhood, and you advertised in the magazines they read and on the TV shows they watched. In an age of virtually indistinguishable products, brands were crucial. This has all changed. "In 50 years," Morgan says, through the darkness, "we've gone from a command-and-control economy to one driven by consumers." How did this happen? For starters, computers made their way into factories. This gave manufacturers new flexibility. It became much easier to tweak cereals or sodas to create nuttier or more lemony blends. With a simple command, looms weaving a striped pattern switched to plaid. This wasn't much harder than it is for me to change the font from Times to Papyrus as I write this chapter. And it meant that industry could produce thousands of new variations. At the same time, globalization was dumping products from all over the world practically on our doorstep. Today, choices are nearly limitless. Winning in this crowded marketplace requires far more than industrial efficiency. The trick now is to deliver to each of us the precise flavor and texture and color we want, at just the right price. Consumers run the show, Morgan says. "It's not controlled by manufacturing or distribution."

This means that marketers must scope us out as individuals. One approach would be to deploy battalions of psychology and lit majors armed with clipboards to knock on our doors. That's impractical. The sensible way to study us is to track and analyze the data we never stop spewing. And Morgan is stretching beyond that. He tells me of experiments his team is developing to monitor the spark of recognition in the brain as people look at online ads. The tests focus on a brain wave called p300. (The U.S. Navy has run similar tests to see how pilots distinguish friends from foes in the air.) If a p300 wave heats up within a fraction of a second of a subject's seeing an ad, the Tacoda team will make the case that the viewer has not only looked at the spot but has processed it mentally. The next step? Figuring out which type of people process certain types of ads. Like other Numerati in a wide range of industries, Dave Morgan is scrutinizing humans and searching for hidden correlations. What do we do, he asks, that might predict what we'll do next?

WHEN I TELL people about this book, they often say, "We're just going to be numbers!"

Yes, I say, but we've long been numbers. Think of the endless rows of workers threading together electronic cables in a Mexican assembly plant or the thousands of soldiers rushing into machine-gun fire at Verdun—even the blissed-out crowd pushing through the turnstiles at a Grateful Dead concert. From management's point of view, all of us in these scenarios might as well be nameless and faceless. We're utterly interchangeable. Turning us into simple numbers was what happened in the industrial age. That was yesterday's story.

The Numerati have much more ambitious plans for us. Forget single digits. They want to calculate for each of us a huge and complex maze of numbers and equations. These are mathematical models. Scientists have been using them for decades to simulate everything from fleets of trucks to nuclear bombs. They build them from vast collections of data, with every piece representing a fact or a probability. Each model must reflect, in numbers, the physical truth: its size and weight, the characteristics of its metal and plastics, how it responds to changes in air pressure or heat. Complex models can have thousands, or even millions, of variables. And they must interact with one another mathematically just the way they do in the real world. Building them is painstaking work. And sometimes they flop. The dramatic market convulsions of 2008, for example, stemmed from faulty models that glossed over the complexity—and the risk—associated with real estate loans.

Despite such stumbles, today's Numerati are plowing forward, with an eye on us. They're already stitching bits of our data into predictive models, and they're just getting warmed up. In the coming decade, each of us will spawn, often unwittingly, models of ourselves in nearly every walk of life. We'll be modeled as workers, patients, soldiers, lovers, shoppers, and voters. In these early days, many of the models are still primitive, making us look like stick figures. The ultimate goal, though, is to build versions of humans that are just as complex as we are—each one unique. Add all of these efforts together, and we're witnessing (as well as experiencing) the mathematical modeling of humanity. It promises to be one of the great undertakings of the twenty-first century. It will grow in scope to include much of the physical world as mathematicians get their hands on new flows of data, from constellations of atmospheric sensors to the feeds from millions of security cameras. It's a parallel world that's taking shape, a laboratory for innovation and discovery composed of numbers, vectors, and algorithms. And you and I are in the middle of it.

What will the Numerati learn about us as they turn us into dizzying combinations of numbers? First they need to find us. Say you're a potential SUV shopper in the northern suburbs of New York, or a churchgoing, antiabortion Democrat in Albuquerque. Maybe you're a Java programmer ready to relocate to Hyderabad, or a jazz-loving, Chianti-sipping Sagittarius looking for snuggles by the fireplace in Stockholm. Heaven help us: maybe you're eager to strap bombs to your waist and climb onto a bus. Whatever you are—and each of us is a lot of things—companies and governments want to identify and locate you. Consider this: Google grew into a multibillion-dollar sensation by helping us find the right Web page. How much more valuable will it be, in every conceivable industry, to find the right person? That information is worth fortunes, and the personal data we throw off draws countless paths straight to our door. Even if you hold back your name, it's a cinch to find you. A Carnegie Mellon University study recently showed that simply by disclosing gender, birth date, and postal zip code, 87 percent of people in the United States could be pinpointed by name.

The Numerati also want to alter our behavior. If we're shopping, they want us to buy more. At the workplace, they're out to boost our productivity. As patients, they want us healthier and cheaper. As companies such as IBM and Amazon roll out early models of us, they can predict our behavior and experiment with us. They can simulate changes in a store or an office and see how we would likely react. And they can attempt to calculate mathematically how to boost our performance. How would shoppers like you respond to a $100 rebate on top-of-the-line Nikon cameras? How much more productive would you be at the office if you had a $600 course on spreadsheets? How would your colleagues cope if the company eliminated their positions or folded them into operations in Bangalore? The Numerati will be placing our models in all kinds of scenarios. They'll try out different medicines or advertisements on us. They'll see how we might respond to a new exercise regimen or a job transfer to a distant division. We don't have to participate or even know that our mathematical ghosts are laboring night and day as lab rats. We'll receive the results of these studies—the optimum course—as helpful suggestions, prescriptions, or marching orders.

The exploding world of data, as we'll see, is a giant laboratory of human behavior. It's a test bed for the social sciences, for economic behavior and psychology. Researchers at companies such as Microsoft and Yahoo are busy hiring scientists from fields as diverse as medicine and linguistics to help them grapple with the bits of our lives that are pouring in. These streams of digital data don't recognize ancient boundaries. They're defined by algorithms, not disciplines. They can easily cross-fertilize. This means that psychologists, economists, biologists, and computer scientists can collaborate as never before, all of them sifting for answers through countless details of our lives. Jack Einhorn, the chief scientist at a New York media start-up called Inform Technologies, predicts that the great discoveries of the twenty-first century will come from finding patterns in vast archives of data. "The next Jonas Salk will be a mathematician," he says, "not a doctor."

IT'S MIDSUMMER gridlock in Manhattan. By the time I reach the French bistro in Chelsea, Dave Morgan's already sitting at a table by an open window, reading e-mails on his Treo. He seems distracted as we eat, glancing from time to time at the handset. Just as the waitress drops the dessert menus on our table, his machine beeps. Morgan looks at it, apologizes, and hurries off into the summer heat. From my seat at the window, I watch him angling across the street and trotting up the far sidewalk.

The next time I see Morgan, it's October. He's moved from Tacoda's Seventh Avenue offices and is newly installed at the headquarters of AOL, high above the skating rink at Rockefeller Center. I meet him at the door of what he calls 75 Rock, and we walk to a café. He tells me that on the day we had lunch, he and his investors agreed to sell Tacoda to AOL. (The reported price was $275 million. The Numerati, it should be noted, tend to make a lot of money.) Morgan is working, at least for the time being, as a senior advertising exec at AOL. He certainly doesn't need the salary. But he says he's tempted to stick around. By tapping AOL's resources and its millions of users, he says, he can learn even more about Web surfers and target us with ever greater precision. It's a long process, he says. "We're just at the beginning."

I ask him about the correlation he told me about earlier, the one between romantic-movie fans and Alamo Rent A Car. It takes a moment for him to recall it. "Oh yeah. They were off the charts." Did his researchers, I ask, ever come up with an explanation for it? He nods. "It had to do with weekends." It was Alamo ads promoting "escapes" that attracted the attention of these Web surfers, he says. The romantic-movie fans booked leisure rentals, largely for weekend getaways. Perhaps they wanted to act out the kinds of scenes that drew them to the cinema. Banners for weekday rentals apparently left them cold.

This brings Morgan to a different insight, one that involves not just who we are but how we feel. No doubt plenty of romantic-movie fans, he says, rent cars for business trips. But after reading the review of the latest candlelight-and-kisses movie, they're thinking about getaways to Napa Valley or Nantucket. Work, at least for the moment, is far away. The challenge ahead is to map not just our tastes and preferences but our shifting moods. "If you think about it," he says, "the movies and music that people click on tell us a lot about their state of mind at that moment. Are they happy? Are they reflective?" He considers the trove of mood messages that pour through our cell phones. That's a new frontier and a potential gold mine of behavioral data. He goes on about the advertising possibilities of music sites, including AOL's, where they can see us clicking on cheerful, sad, or inspirational songs.

I'm not so sure about that. If I click on a happy song, I say, maybe I'm just looking for a pick-me-up. Morgan shrugs. He won't know until he does more research. This means more of our data to collect and more numbers to run. Just thinking about it makes him smile. Outside the sky grows dark, and rain scatters the crowd at Rockefeller Center. As Dave Morgan heads back to his behavioral laboratory at 75 Rock, he covers his head with his hands and sprints.

CHAPTER 1

Worker

IT'S RUSH HOUR in New York. I stop by Hank's stand on 47th Street, spend a buck and a quarter for a sweetened coffee, carry it to the elevator, and ride high up in a Midtown skyscraper. A big pile of *Wall Street Journals* used to wait at reception, one for each of us. No more. Now we've been instructed to read the paper online. With that, even more of our work moves onto the computer.

I pry the lid off the coffee. I call up Yahoo, read my personal mail, and type a quick reply to an e-mail from my sister. Then I check the Philadelphia papers for baseball news. The Phillies got crushed . . . It's 10 A.M., the coffee's a brown stain on the bottom of the cup, and I'm just getting to the *Wall Street Journal* online. Or maybe I'm not.

Office workers have had pleasant little stalling routines forever, and it hasn't mattered much. Other laborers haven't been so lucky. A century ago, men carrying notebooks and stopwatches made their way into factories and started to measure the movements of workers. They turned industrial production into a science, which reached its zenith in Japanese auto plants. They perfected Statistical Quality Control, and today they can analyze each spray gun, each furnace, and, by extension, each worker, minute by minute. If any one of these elements is missing a beat, they can adjust it on the spot. Many office dweebs, by comparison, luxuriate in privacy. Unless we happen to be snoring louder than usual in our cubicle when the boss strolls by, our work habits remain our own little secret. We're scored on results, not process. Sell a house, win a trial, wow the boss with elegant lines of software code, and we're golden.

Things are changing, though. In the past decade, much of the work we do has moved away from the piles on our desks, the notebooks and newspapers and Post-its stuck to the door. It has migrated right onto the computer, which is now linked to a network. We're tied to a workmate equipped with a phenomenal memory, an uncanny sense of time, and no loyalty to us. He works for the boss, who can measure our efforts with no need for a notebook or a stopwatch. The computer will rat on us, exposing each one of our online secrets without a nanosecond of hesitation or regret. At work, perhaps more than anywhere else, we are in danger of becoming data serfs—slaves to the information we produce. Every keystroke at the office can now be recorded and mathematically analyzed. We don't own them. If our bosses wanted to, they could order up an e-mail chart for each of us. It would display the words we write most often, in proportionally sized fonts. You could only pray that *movies* or *beer* wouldn't show up bigger on your chart than the medicines you sell or the stocks you recommend. That online version of the *Wall Street Journal*? Our employers can follow which articles we read. They can also buy software to create maps of the people we communicate with—our social networks. From these, they can draw powerful conclusions about our productivity, our happiness at work, and our relations with colleagues. Just what kind of team player are you, anyway? Microsoft even filed in 2006 to patent a technology to monitor the heart rate, blood pressure, galvanic skin response, and facial expressions of office workers. The idea, according to the application, is that managers would receive alerts if workers were experiencing heightened frustration or stress. Such systems are in the early stages of research. But even with today's technology, if your company is not scouring the patterns of your behavior at the keyboard, it's only because it doesn't choose to—or hasn't gotten around to it yet.

Why would companies intrude like this? Very simply, to boost our productivity. For centuries they've concentrated on results because, like the newspaper advertisers now rushing to Dave Morgan's offices at Tacoda, they haven't had the means to monitor and dissect what we actually do. Now the tools are at hand. Don't they have a responsibility to shareholders to put them to use and pump up productivity and profits? That's the way they see it.

Now as I look at the workplace through their purposeful eyes, I'm already feeling a trace of nostalgia for the idle moments and wasteful routines that brighten my days. Sitting in my 43rd-floor office, I call up YouTube and click on a silly Morphing Pug video. An animated dog dances and sings a ridiculous song. I wonder what that investment of 45 seconds of utter nonsense could possibly say to my bosses about me. Is there a correlation between Morphing Pug watchers and prizewinning journalism? It's doubtful. And it's a matter of time before management starts recording such behavior. The very thought fills me with such regret that I click on the video once more, not so much to laugh at the dog as to soak up the on-the-job freedom it represents.

On a late spring morning I drive over the Tappan Zee Bridge, across the wide expanse of the Hudson. Then I hook left, away from New York City and up into the forests of Westchester County, to the headquarters of IBM's Thomas J. Watson Research Laboratory. It sits like a fortress atop a hill, a long, curved wall of glass reflecting the cotton-ball clouds floating above. I have a date there with Samer Takriti, the Syrian-born mathematician who launched me on this entire project. He was the one who described to me early on how his team was building mathematical models of thousands of IBM's tech consultants. The idea, he said, was to piece together inventories of all of their skills and then to calculate, mathematically, how best to deploy them. I came away from that meeting convinced that if Takriti could model people as workers, then eventually we'd also be modeled as shoppers and patients—in short, in a whole range of our activities as humans. Now I'm going back to find out how Takriti and his team plan to turn IBM's workers into numbers—and what they'll do with them (and with us) if they succeed.

Takriti, a slim 40-year-old with wide, languid eyes, opens the door of his small office. He wears a rugby shirt tucked tightly into blue jeans. He's on a conference call but waves me in. On one wall of his windowless office is a whiteboard covered with math calculations that mean nothing to me. Takriti is quiet on the call, just saying, "A hum, a hum." I look to the other wall, which is decorated with an electricity grid of New York and Pennsylvania. This is an artifact from Takriti's previous life, when he used math to model chunks of the old economy, things like steel mills and power plants. Story has it, Takriti says after he hangs up, that the original Takritis were warriors who marched from Saddam's native city, Tikrit, in Iraq. His branch of the family, he tells me, eventually settled in Syria. A top engineering student in Damascus, Takriti won a fellowship in the mid-1980s to study at the University of Michigan. He fell head over heels for math. In 1996, by then a Ph.D., he landed a research job at IBM's fabled Watson Research Center, a half-hour drive north of New York City. This son of Tikrit warriors now walked among the gods of math.

Takriti's specialty was stochastic analysis. This is the math that attempts to tie predictions to random events. Say it rains in Tucson from zero to six times per month, and you listen to the weather report, which has been right 19 of the past 20 days, only three times a week. One of your three jackets is suede. What are the chances it'll get drenched tomorrow? Imagine that same question with one thousand variables, and you've stepped into the stochastic world.

A generation ago, a crew of math whizzes led by Myron Scholes and Fischer Black focused their mastery of probability on finance, where they calculated risk and put prices on it. This led to a panoply of new financial products, from options to hedging strategies. It was a math revolution on Wall Street. The mathematicians were replacing hunches, wholesale, with science. Takriti says that by the time he reached IBM, many of the same math tools were being refitted for other industries.

Like energy. Takriti doesn't like to broadcast it, but he left Big Blue in 1999 for Houston, where he worked for Enron. Back then, Enron was not only innovating the kind of corporate fraud that would lead to its collapse. It also ran a world-class mathematics laboratory. The entire world, as Enron saw it (and was soon to demonstrate all too vividly) was awash in uncertainty. People had trillions of dollars riding on chance. If you looked at weather as a topsy-turvy market, for example, theme parks were betting on sunshine, farmers on rain. Enron's math team could calculate the weather risks and then develop indexes and financial options for cold fronts and dog days. Everyone could hedge the weather, and Enron would turn this into a business. Given enough mathematicians, it seemed, every chancy element in the world could eventually be quantified, modeled, and turned into a financial instrument.

Takriti's stock soared at Enron. And when IBM called in late 2000, they offered him the top job in stochastic analysis. Takriti jumped. He got out of Houston, it turned out, barely a year before Enron's collapse. His new focus at IBM would be every bit as hard to quantify and predict as flash floods in the Mojave Desert or the looming corporate bankruptcy in Houston. Takriti would be modeling human workers.

I tell Takriti that being modeled doesn't sound like much fun. I picture an all-knowing boss anticipating my every move, perhaps sending me an e-mail with the simple message, "No!" before I even get up my nerve to ask for a raise. But Takriti focuses on the positive. Imagine that your boss finally recognizes your strengths, he says — maybe ones that are hidden even to you. Then he "puts you into situations where you will thrive."

If your performance is stellar, companies eventually could wield your mathematical model as a specimen of workplace DNA. And they could use it, in a sense, to clone you. Imagine, says Aleksandra Mojsilovic, one of Takriti's modelers, that the company has a superior worker named Joe Smith. Management could use two or three others just like him, or even a dozen. Once the company has built rich mathematical profiles of their employees, it shouldn't be too hard to sift through them to identify the experiences or routines that make Joe Smith so good. "If you had the full employment history, you could even compute the steps to become a Joe Smith," she says. Most of this, of course, would involve training programs, not genetic manipulation. And the real Joe Smith may have intuitive smarts or a knack for design that just cannot be replicated. "I'm not saying you can re-create a scientist, or a painter, or a musician," Mojsilovic says. "But there are a lot of job roles that are really commodities." And if people turn out to be poorly designed for these jobs, they'll be reconfigured, first mathematically and then in life.

When Samer Takriti sits down to define one of his colleagues in symbols, he looks to economists and industrial engineers for guidance. They've been modeling complex systems for decades. Economically, he regards us as components in a labor market. Our value rises and falls with demand. In that sense, we fit into the financial equations developed on Wall Street. And when we're employed, what do we do? We work with colleagues to build things and create value. So, boiled down to numbers, we share at least a few mathematical properties with the components that are unloaded every day at IBM's huge microprocessor factory up the road, in Fishkill, New York. Look at us one way, and we're stocks. Change the perspective, and we're machine parts.

Of course, this isn't entirely fair. We're more than stocks and parts, quite a bit more. Takriti is the first to admit it. It's because we're so different—so hard to predict—that Takriti needs a team of 40 Ph.D.'s, from data miners to linguists, to decode our behavior and our traits. They catalog what they find—each of our gestures, each of our skills—into symbols that a computer can digest. "Everything must be turned into numbers," Takriti says.

One of Takriti's challenges is to help IBM develop a taxonomy of the skills of its 300,000 employees. On its balance sheet, IBM lays out the value of many other assets, from su-

percomputers to swiveling Aeron desk chairs. When strategists at the company are figuring out whether to sell a division or invest more money, they pore over these figures. They
sketch out rosy and grim scenarios. They do the numbers.

But how do they "do the numbers" on you and me? Yes,
they know how much we cost. Anything that's counted in
currency fits neatly into their equations. But what do they
get for that money? How can that be measured? What's our
potential? Will there be a glut of people like us in the next
few years? A shortage? Planners want answers. To carry out
these calculations, they have to turn us into something that,
like financial instruments, can be measured over time. Picture an average worker in an industry that's chugging along
at its usual pace. At the risk of seeming cold-hearted, let's give
that imaginary laborer a rank based on his current value. Call
him a C. If the industry heats up and more such workers are
needed, his value rises, maybe to C+ or even B. If he picks up
more skills or starts working harder, the same thing happens.
His stock rises. But if the industry plunges into recession and
companies shut down operations, our worker finds himself in
a surplus market. His stock plummets, down to a D or even
an F. We're all too familiar with this dynamic. Workers find
jobs in boom times and get laid off in slumps. But often the
process has little to do with a worker's value. In some companies, the last workers hired are the first to get the boot. That
rewards longevity, not value. Sometimes it's the friendly workers who survive, or even workers who have a knee-breaking
cousin in the mob. These are metrics a caveman could grasp.
The Numerati have a different plan altogether. But how will
they calculate our worth? How will they turn us into quantifi-
able financial instruments?

The first step is to break us down into little pieces. These
are the characteristics we share with others, the bits of us that

can be squeezed into columns and assigned numbers. Computers, after all, aren't yet capable of appreciating us as the integrated and complex beasts that Leo Tolstoy wrote about. You might have the nicest smile on earth and wonderful rapport with colleagues. Maybe you're mean, or smell like onions. There's no room, at least in the early versions of IBM's employee database, for those personal details. Some of them may be crucial. They may represent the real you. But the database understands us largely as a mosaic of résumé items, from job categories to mastery of the computer language C++ to fluency in Mandarin.

It's pathetically shallow. Consider what happens when you sit down in a room to, say, hammer out a new marketing campaign with five colleagues. This is life in the analog world. Your brain, by far the most sophisticated computing device known in the universe, processes an astonishing range of data. It perceives a wrinkled nose, a sideways glance, a hint of sarcasm, a flash of disdain. It ties together smells and sounds, and it links them with other memories and lessons from the past. Add up all the words and looks and gestures, and your brain picks up thousands, or even millions, of signals emanating from those five people. In his book *Strangers to Ourselves,* Timothy Wilson of the University of Virginia notes that as data streams in from our five senses, the brain grapples with more than 11 million disparate pieces of information per second. Today's computers cannot handle such complexity. IBM's mathematical system may scan each of us for a mere five or ten data points. I've had dogs that dig deeper into human nature. However, once we're represented as bits of math, the machine can do something superhuman. It can mix and match us in a fraction of a second with a million, or 100 million, others. That scale promises new efficiencies—and insights.

Imagine what IBM's bean counters will be able to do

once all of the company's workers are classified by their skills. They'll start running ever more detailed numbers on workers —just as they do on other investments. They'll attempt to calculate the financial return for each job category and each skill, whether it's Java programmers or office managers. They'll compare productivity in ever-greater detail, worker by worker and region by region. This will help them decide which jobs to send offshore. And they'll be able to measure productivity based on dozens of yardsticks. How productive are workers in your category as they reach ages 45, 50, and 60? Once the company has those numbers, they might be able to calculate not just the present value of workers but also what they'll be worth down the road.

This will take some getting used to. To date, we've managed our human relations in an old-fashioned economy, one largely lacking in numbers and metrics. For the most part, whether we're looking for a favor or even a mate, we've bartered: here's what I'll provide you; here's what I want in return. Nothing to measure, little to count. For centuries, even commerce worked this way. You won't give me those two goats for this table? How about if I throw in a hammer? This process is painfully inefficient. Each barter calls for more haggling. Values fluctuate. No surprise, then, that bartering retreated into the hills as soon as societies came up with a numeric symbol for value—money. This was a triumph of the earliest Numerati. It provided a math tool to count and calculate and compare a world of different things. And it eventually led to expanded commerce, global markets, the numbers blazing on the liquid crystal screens of the Tokyo Stock Exchange. Now Takriti and his team are turning us into symbols so that we can take our place in new human markets.

Just the way brokers run portfolios of junk bonds or

emerging market stocks, the Numerati are dropping us into portfolios of people. It's happening in industry after industry. Alamo Rent A Car, for example, buys a portfolio of romance-movie lovers from Tacoda and then compares its performance to that of others. If Takriti and his team manage to reduce IBM's work force into a coherent portfolio of skills—something a computer could understand—IBM could soon deploy its labor in much the way that it manages its financial investments. That's precisely what Takriti has in mind.

This has been the case for years in baseball. On my shelf I have a fat encyclopedia listing baseball statistics on every major league player since the 1880s. But now the baseball Numerati are slicing and dicing the data nearly as fast as the Wall Street quants (quantitative analysts, that is). They're coming up with new metrics—new ways to model the players mathematically. One new statistic is called WARP, "wins above replacement player." On a quant site called Baseball Prospectus, I'm looking at a profile of Carlos Beltran, the elegant switch hitter who patrols center field for the New York Mets. In late 2004, he signed a seven-year contract for $120 million, or an average of $18 million a year. After a stellar 2006 season, his WARP stood at 10.6. That means that if the Mets swapped him for a commodity replacement player making a laughable half million a year, the team would win nearly 11 fewer games in a 162-game season. Each additional victory, by these numbers, would cost the Mets $1.62 million—an affordable price for a rich New York team. But will Beltran's WARP be nearly as high in 2010, when he's a creaky 33 years old? Sadly, Baseball Prospective says no. It predicts that Beltran's WARP by the decade's end will decline to 3.6 and that he'll be worth only $5.8 million—far less than he'll earn.

Is that calculation right? It's anybody's guess. Does the

WARP account for Beltran's intangibles, all those immeasurable qualities—the helpful batting tips he gives a rookie or the way he distracts the other team's pitcher as he dances off first base? In other words, do the numbers reflect reality in all of its complexity? Often they fall short, even in the statistically exuberant realm of baseball. Pick the wrong numbers, and they can lie. That's no secret. But just try making that argument to your boss when your numbers slump.

I confess to Takriti that I find the prospect of being scored like Carlos Beltran a bit unsettling. It's nice, I've found, to live and work off the database, in the foggy barter economy. True, it's a real headache for bean counters. But the unmeasured universe can be a forgiving place. Smiles, friendships, and even artfully spun yarns all count for something there—maybe a smidgen of job security or even a raise. Workers in a measured workplace are on their own and more likely to rise or fall with their numbers. Precious few of them have seven-year contracts like Carlos Beltran's. And for these quantified masses, the security of the flock fades away. After all, each lazy or incompetent worker who survives in the mathematically assessed workplace represents a market inefficiency. Once the measurements are in place, these workers will presumably plunge in value or be purged, just like an underperforming stock in a portfolio.

Think you can manage life in a portfolio of workers? It could be the best thing that ever happened to you. Some stocks shoot through the roof, and certain workers will too. But being dropped into a portfolio is only the beginning. Takriti and his team are already building the next stage, in which we'll be understood and interpreted in far greater detail.

· · ·

ON A NOVEMBER morning in 2006, IBM's chairman, Sam Palmisano, stepped up to a podium in the Forbidden City, in the heart of Beijing. He had an announcement to make. Palmisano was dressed in his standard business suit and wore his trademark horn-rimmed glasses. But as he made his way to the podium, something about him looked not quite real, more like a cartoon. This was not the real Palmisano, it turned out, but an avatar representing him. And the Forbidden City he visited was a digital simulation. IBM technicians had built it and mounted it within the virtual world called Second Life. Journalists who wanted to hear Palmisano's announcement grumbled about this venue for weeks. They had to sign up for Second Life and attend as their own avatars.

By staking IBM's blue flag in a simulated world, Palmisano was pointing to the company's future. Already, engineers around the world use computer simulations to design electric turbines and fine-tune the traffic flows in major cities. The way IBM sees it, entire business processes will one day be simulated. Picture managers, their fists grasping joysticks, trying out new industrial approaches and calibrating operations as if they were running their own version of the video game The Sims. If Takriti and his team master their next assignment, the avatars on the screen will be the mathematical models of IBM's workers.

This process is just beginning, and Takriti already waxes nostalgic for the old days, when it was machines that were modeled. They're simpler. Machines don't cheat, feud, pout, develop serious drinking problems, or get depressed. And they don't come up with great and transformative ideas. Takriti goes on for a while about the maddening randomness of humans.

I interrupt to ask about the math involved. I point to the

whiteboard covered with formulas and notations, some of them snaking up and down to make room for others. "What are you working on here?" (Some of these notations are new to me.)

He shrugs. Takriti, like many Numerati, tends to downplay the complexity of the formulas he scribbles so effortlessly. He rejects the notion that he and his confreres draw their algorithms and equations from a magical toolbox. Part of this is modesty. But Takriti also has the conviction that even the nitty-gritty of stochastic calculus would be clear to outsiders if we just sat still and paid attention. He starts to explain one of the formulas. Then he stops. It's the humans that are hard to figure out, he insists. "Math is the easy part."

For decades, IBM researchers have been transforming bigger and bigger pieces of the company's business into math. The science they use, known as operations research, was born during World War II. German submarines, known as U-boats, were attacking convoys and sinking lots of ships. How should the convoys be deployed, the mathematicians were asked, to minimize the damage? Was it better to travel in large groups, escorted by many destroyers? Or would smaller ones be harder for the U-boats to track down?

Math whizzes at the U.S. Antisubmarine Warfare Operations Research Group (ASWORG) built mathematical representations of the convoys. These were models, and they operated within a set of constraints—conditions imposed by the real world. The ships couldn't move faster than a certain speed, for example, and they had to carry enough food and fuel to reach their destination. They had to steer clear of icebergs. The mathematicians also had statistics on the U-boats—the size of the fleet, the range of the subs, the deadliness of their missiles. Using this information, they could model the naval war. Each vessel was linked to others by numbers, by the probability that something good, bad, or indifferent would happen to it. These fleets in the North Atlantic existed in their model as a web of statistical relationships. As the researchers tinkered with the fleet in the model, the odds changed. The ASWORG team was able to calculate that large convoys with big escorts were significantly safer. They determined how far down to send depth charges to inflict the most damage on enemy subs. As the U.S. Navy put these formulas into practice, the attrition of the convoy ships dropped. Shipments reached Britain. By the end of the war, mathematicians were using similar methods to boost the efficiency of anti-aircraft defenses and fuel depots.

Takriti is telling me about one of the giants of the field, George Dantzig, when he jumps to his feet, reaches high on a shelf for a big textbook, and starts flipping through it. "Dantzig did the mathematics of marriage," he says. "Maybe it's something you can use in your dating chapter." Dantzig, I learn, regarded multiple sexual partners as variables, and he attempted to prove that monogamy—at least from the coolheaded perspective of an operations researcher—yielded better results than polygamy. Takriti doesn't find the details in the book. Maybe I could find it on the Web, he says. That turns out to be a cinch, though I think it's safe to say that Dantzig's study, while fascinating to the Numerati, left the institution of marriage largely untouched.

But outside of matrimony, his influence is with us every day. In 1947, the Berkeley-trained mathematician came up with the so-called simplex algorithm. An algorithm is nothing more than a recipe—an ordered set of commands. This one was a recipe to guide intelligent decision making. If farmers wanted to know which type of seed to plant in a particular soil, or if steelmakers wondered whether to haul coal on trucks or barges, operation researchers had answers. They just needed the numbers, the constraints, and the goal. Using Dantzig's algorithm, they could find the point where the objective, whether it involved dollars or tonnage, reached its zenith, its optimum point. Then they could calculate, working backward, how to come up with that result. Known as optimization, this process now guides logistics and planning and network design in much of the modern world. If you want to fly from Los Angeles to New York, Travelocity's optimization program flips at lightning speed through 10,000 possible routes and finds the one that will make the most sense for you and the most money for Travelocity and its partners (profit is one of its constraints). Military planners optimize helicopter routes over insurgent hotbeds in Iraq. And when you make a call on your cell phone, an optimization program chooses the best pathway of towers to convey the signal.

Back when Dantzig was putting the final touches on his algorithm, IBM researchers were already preparing to apply operations research to their own business. They had the mother of all tests for it: IBM's massive supply chain. To build its renowned office machines (which didn't yet include commercial computers), IBM bought parts and raw materials from suppliers all over the world. Naturally, these were a major expense. If the company could use this new math to organize it all, the savings would drop straight to the bottom line.

The math worked. In fact, IBM was able to turn this particular know-how into a business. The company's experts helped other companies convert their own logistics into math and then optimize them. This is where the story turns inside out, a bit like that drawing by M. C. Escher, where the artist's hand is drawing itself. In the past couple of decades, IBM's focus moved from manufacturing to services. The company now sells more expertise than machinery. It unloaded its personal computer division to China's Lenovo in 2005, and IBM Global Services has grown into a $40 billion business. So if IBM's experts were to optimize their supply chain today, they would have to model and fine-tune themselves. That's precisely what Takriti's team is busy doing.

Just think where this could lead. We've seen, with supply chains, how the company used itself as a laboratory. It mastered the process for itself and then sold the expertise to others. Now the company is modeling its workers. If this leads to big gains in productivity, do you think that expertise will remain locked up inside Big Blue? I don't. Imagine mathematical modelers arriving at the doors of your company one day, either as a phalanx of blue-clad consultants or perhaps encoded in a piece of software. Their focus will be on you.

SITTING IN HIS small office, one blue-jeaned leg crossed over the other, Samer Takriti confesses to me that he's nervous. I can't blame him. His assignment is to construct detailed mathematical models of 50,000 of his colleagues. We're not talking about simply placing workers and their jobs into the kind of bare-bones taxonomy we described earlier. That's complicated enough. The goal here is to build entire models, complete with a person's quirks, daily commute, and allies and enemies. These models might one day include whether the workers eat beef or pork, how seriously they take the Sabbath, whether a bee sting or a peanut sauce could lay them low. No doubt, some of them thrive even in the filthy air in Beijing or Mexico City, while others wheeze. If so, the models would eventually include this detail, among countless others. Takriti's job is to depict flesh-and-blood humans as math.

Takriti is not given to bold forecasts. But if his system is successful, here's how it will work: Picture an IBM manager who gets an assignment to send a team of five to set up a call center in Manila. She sits down at the computer and fills out a form. It's almost like booking a vacation online. She puts in the dates and clicks on menus to describe the job and the skills needed. Perhaps she stipulates the ideal budget range. The results come back, recommending a particular team. All the skills are represented. Maybe three of the five people have a history of working together smoothly. They all have passports and live near airports with direct flights to Manila. One of them even speaks Tagalog. Everything looks fine, except for one line that's highlighted in red. The budget. It's $40,000 over! The manager sees that the computer architect on the team is a veritable luminary, a guy who gets written up in the trade press. Sure, he's a 98.7 percent fit for the job, but he costs $1,000 an hour. It's as if she shopped for a weekend getaway in Paris and wound up with a penthouse suite at the Ritz.

Hmmm. The manager asks the system for a cheaper architect. New options come back. One is a new 29-year-old consultant based in India who costs only $85 per hour. That would certainly patch the hole in the budget. Unfortunately, he's only a 69 percent fit for the job. Still, he can handle it, according to the computer, if he gets two weeks of training. Can the job be delayed?

This is management in a world run by Numerati. As IBM sees it, the company has little choice. The work force is too big, the world too vast and complicated for managers to get a grip on their workers the old-fashioned way—by talking to people who know people who know people. Word of mouth is too foggy and slow for the global economy. Personal con-

nections are too constricted. Managers need the zip of automation to unearth a consultant in New Delhi, just the way a
generation ago they located a shipment of condensers in To-
peka. For this to work, the consultant—just like the condensers—must be represented as a series of numbers.

To put together these profiles, Takriti requires mountains
of facts about each employee. He has unleashed a squadron
of Ph.D.'s, from data miners and statisticians to anthropologists, to comb through workers' data. Personnel files, which
include annual evaluations, are off limits at IBM. But practically every other bit of data is fair game. Sifting through résumés and project records, the team can assemble a profile
of each worker's skills and experience. Online calendars show
how employees use their time and who they meet with. By
tracking the use of cell phones and handheld computers,
Takriti's researchers may be able to map the workers' movements. Call records and e-mails define the social networks of
each consultant. Who do they copy on their e-mails? Do they
send blind copies to anyone? These hidden messages could
point to the growth of informal networks within the company. They may show that a midlevel manager is quietly leading an important group of colleagues—and that his boss is
out of the loop. Maybe those two should switch jobs.

The interpretation of our social networks is an exploding
field of research, from IBM to the terror-trackers at the National Security Agency in Fort Meade, Maryland. One leading lab is at Carnegie Mellon University, in Pittsburgh, where
a professor named Kathleen Carley is building an entire social
network empire within the computer sciences department.
When I meet with Carley, she has 30 grad students jammed
into just a handful of basement offices. They're analyzing the
networks of contagious diseases, such as Asian flu. They're

comparing the dynamics of different networks in the Middle East.

What can this social network analysis divulge about workers at IBM or elsewhere? Lots. Start with e-mail. Carley's grad students can feed a computer all of a company's e-mails over a certain period of time. They practice with the e-mails exchanged during the frantic dying months of Enron. Released as evidence in Enron trials, they've been dissected ever since by social network researchers around the world. Carley's system notes the senders of e-mails, the time the messages were sent, and the recipients. Without even reading the content of the e-mails, a software program her team has built draws various diagrams of the organization. One of them shows who communicates with whom. When she shows it to me, it looks at first like a spaghetti cook-off. The organization—if you can call it that—features different tangled piles, each one with its own set of meatballs. Single noodles extend from one pile to the others. Each meatball, naturally, represents a person within the organization, and the piles represent groups of people who communicate heavily with one another.

Logical enough. The finance people, the gas people, the legal team, they all communicate within their groups, with an occasional e-mail to another department. But it's not that simple. "See this group here?" Carley says, pointing to a cluster of meatballs in a swirling mass. It's an informal network, she says. It took shape as Enron collapsed. This group sent out about one thousand messages a day and became a clearinghouse for inside dope. If the company had been studying this network, executives might have interpreted it as an insurgency taking shape. In a sense, it was, since a growing network of employees was trading ever more dire reports and rumors about a company in crisis—and helping one another prepare for life after Enron.

Corporations elsewhere, including IBM, can draw all kinds of insights from their employees' networks. They can map each person's circle of contacts. They can also spot outliers, people who aren't communicating much with anyone. These employees, Carley says, are worth scrutinizing: they may be depressed or about to leave, or even consorting with the competition. Even without reading all the e-mails, the company can automatically spot the most common words that circulate within each group. This permits them to map not only each worker's contacts but also the nature of those links. They can also see how communications shift with time. Two workers may discuss software programming Tuesday through Friday but spend much of their time on Monday sending e-mails about the past weekend's football games. "The next big step," Carley says (a bit ominously), "is to take tools like this and tie them to scheduling and productivity programs." I read this to mean that we office workers are well on our way to being optimized.

Sound scary? It may depend on where you're perched on the food chain. Remember the $1,000-per-hour consultant who almost got dispatched to the Philippines? He didn't end up going, and instead, in IBM's scheme, he remained "on the bench." Takriti smiles. "That's what we call it," he says. "I think the term comes from sports." The question, of course, is how long IBM wants to have that high-priced talent sitting on the bench. If there isn't any work to justify his immense talents, shouldn't they put him on something else, just to keep him busy?

Not necessarily, says Takriti. Job satisfaction is one of his system's constraints. If workers get angry or bored to tears, their productivity is bound to plummet. The automatic manager keeps this in mind (in a manner of speaking). As you might expect, it deals very gently with superstars. Since they make lots of money for the company during short bursts of activity, they get plenty of time on the bench. But grunt workers in this hierarchy get far less consideration. They're calculated as "commodities." Their skills are "fungible." This means these workers are virtually indistinguishable from others, whether they're in India or Uruguay. They contribute little to profits. It pains Takriti to say this, because humans are not machines. They have varying skills and potential to grow. He appreciates this. But looking at it mathematically, he says, the company should keep its commodity workers laboring as close as possible to 100 percent of the time. Not much kickback time on the bench for them.

Where is this all leading? I pose the question one afternoon to Pierre Haren. A Ph.D. from Massachusetts Institute of Technology and a prominent member of the Numerati, he's the founder and chief executive of ILOG. It's a French company that uses operations research to fine-tune industrial systems, charting, for example, the most efficient delivery routes for Coors beer. ILOG makes allowances for all kinds of constraints. For example, a few years ago, the Singapore government wanted to avoid diplomatic spats at its new airport. So officials asked ILOG to synchronize the flow of passengers, making sure that those from mainland China wouldn't cross paths with travelers from Taiwan. Haren speaks in a strong French accent. We're talking in the lobby of a Midtown hotel in New York, and he has to yell to make himself heard over a particularly loud fountain.

Haren says that the efforts underway at places like IBM will not only break down each worker into sets of skills and knowledge. The same systems will also divide their days and weeks into small periods of time—hours, half-hours, eventually even minutes. At the same time, the jobs that have to be done, whether it's building a software program or designing an airliner, are also broken down into tiny steps. In this sense, Haren might as well be describing the industrial engineering that led to assembly lines a century ago. Big jobs are parsed into thousands of tasks and divided among many workers. But the work Haren is discussing is not done by hand, hydraulic presses, or even robots. It flows from the brain. The labor is defined by knowledge and ideas. As he sees it, that expertise will be tapped minute by minute across the world. This job sharing is already starting to happen, as companies break up projects and move big pieces of them offshore. But once the workers are represented as mathematical models, it will be far easier to break down their days into billable minutes and send their smarts to fulfill jobs all over the world.

Consider IBM's superstar consultant. He's roused off the bench, whether he's on a ski lift at St. Moritz or leading a seminar at Armonk. He reaches into his pocket and sees a message asking for ten minutes of his precious time. He might know just the right algorithm, or perhaps a contact or a customer. Maybe he sends back word that he's busy. (He's a star, after all.) But if he takes part, he assumes his place in what Haren calls a virtual assembly line. "This is the equivalent of the industrial revolution for white-collar workers," Haren says.

Some of us like to think that our work is too creative to be measured and modeled. I used to feel this way. For years I would write articles, and the only metric that mattered was whether the editor in chief appreciated them. Things began to change when the articles moved online. This made it possible for managers to count how many people read each article. Some managers these days rank writers by page views or how many times each article is e-mailed by readers. Is this fair? Not in my view. I remember one time a colleague posted on his blog a video ad featuring Paris Hilton. She wasn't wearing much, and she was washing a car with a big wet sponge in a splashy and provocative fashion. His blog attracted tens of thousands of visits that day, more than others of us got in a month. Did he outperform us? It depends on what the bosses decide to count. As the Numerati gain sway in the workplace, such questions are bound to rage.

It's getting late in Takriti's office. I can see that he's concerned about my line of questioning. This virtual assembly line sounds menacing. The surveillance has more than a whiff of Big Brother. For those of us who aren't Carlos Beltran or a $1,000-per-hour consultant, life as a mathematical model is sounding like abject data serfdom.

Here's Takriti's counterargument. As the tools he's building make workers more productive, the market will reward them. (So there's an economic benefit, even to us serfs.) What's more, workers will increasingly use their numbers to open doors for themselves. We already use math programs to map our trips and look for dates. Why not use them to map our careers—and negotiate for better pay? Let's say analytical tools show that a consultant's value to the company topped $2 million one year. Shouldn't she have access to that number and be free to use it as a negotiating tool? In a workplace defined by metrics, even those of us who like to think that we're beyond measurement will face growing pressure to build our case with numbers of our own.

CHAPTER 2

Shopper

THE CALL COMES from my wife at the supermarket. "Do we need onions?"

I check. "We have one big one," I say, turning it over gingerly. "But it's been sprouting for a while . . ."

"Okay, I'll get some. How about milk?"

You know the routine. A few minutes later, whichever one of us is shopping arrives at the checkout counter. There, if we remember, we dig into a pocket or purse for the frayed customer loyalty card on the key chain. The cashier scans it. We get a discount on the orange juice or razor blades, and the supermarket learns about everything we buy. It's a deal we shoppers have been making for years. Stores give us what amounts to a couple bucks a week in exchange for our shopping lists.

Here's the strange part. To date, retailers have stockpiled untold mountains of our personal data, but they're only now waking up to what they can do with it. Sure, managers have used the scans to keep an eye on inventory. They can see when to order more mangoes or Snickers bars. They've learned plenty about our behavior en masse but next to nothing about us as individuals. When we walk into a store, even if it's the hundredth time this year, the system doesn't recognize us. It's clueless.

This era is coming to an end. Retailers simply cannot afford to keep herding us blindly through stores and malls, flashing discounts on Pampers to widowers in wheelchairs and ham hocks to Jews who keep kosher. It's wasteful, and competitors are getting smarter. Look online. Whether it's Amazon.com or a travel service like Orbitz, Internet merchants are working every day to figure us out.

They're tracking every click on their sites. They know where we come from, what we buy, how much we spend, which advertisements we see. They even know which ones we linger over for a moment or two with our mouse. In the online world, businesses no longer look at us as herds but as vast collections of individuals—each of us represented by scores of equations. They prove every day that merchants who know their customers have a big edge. They can study our patterns of consumption, anticipate our appetites, and entice us to spend money.

Personal service is nothing new for retailers. For centuries, it's been a privilege for the rich. Shopkeepers and tailors know their names and measurements and their taste in premier cru burgundies. They also know where to send the bill. A few generations ago, the rest of us got personal service (on a far more modest scale) in our own neighborhoods. "The retail model was a shopkeeper, a millinery, a rug merchant," says Jeff Smith, a managing partner of the retail practice at Accenture, the tech consulting giant. "You didn't serve yourself," he says. "They stood behind counters and found what you were looking for." Chummy relations with customers gave these merchants an edge.

Following World War II, however, retail took a half-century detour into mass industrialization. Shoppers were handed carts and instructed to find their own stuff. Whether they were pushing those carts through Ikea or Wal-Mart, they had entire warehouses to explore. And the merchandise was cheap, in part because the stores had eliminated the middle-man—the shopkeeper at the local store who knew the customers by name. They mastered a startling new efficiency, which came from manufacturing and distributing with martial precision. That's what the brainiacs and their computers were focused on: operations. The customers? As we made our way from the massive lots through the equally massive stores, we were processed like card-carrying herd animals.

Now retailers are changing. Accenture's Smith calls it "back to the future." Instead of deploying millions of shopkeepers to twenty-first-century counters, they're relying on automatic machinery, from video cameras to newfangled customer loyalty cards. The operation runs on data, our data. The goal is to follow our footsteps in much same the way that e-tailers track our clicks. In the marketplace of the Numerati, we'll define ourselves as shoppers in ever-greater detail simply by going about our business in a store. When the stores get to know us, they'll recognize us the moment we walk in the door—just the way the corner grocer used to. And just like that grocer, they'll know our week-to-week routines and our not-so-secret cravings. They may calculate that we're probably running short on cat kibbles, and they won't forget that we spike a gallon or two of eggnog every holiday season. (And wouldn't it taste better with premium Jamaican rum this year?) The automatic systems will calculate not only what we're likely to buy but also how much money we make for the store. Many of them will learn how to lavish big spenders with special attention and nudge cheapskates toward the door.

· · ·

AN OLD shopping cart is parked next to the wall at Accenture's lab, high above downtown Chicago. The offices are chock full of tech gadgetry. Blinking video cameras hang from the ceilings, staring down on the researchers. (They're guinea pigs in a new surveillance system designed to track shoppers and workers.) In one nook of the lab is a large, always-on video connection with another Accenture lab in Silicon Valley. Around lunchtime in Chicago, you can see the California contingent coming to work, steaming coffee cups in hand. You hear their phones ringing and their footsteps echoing across the lobby 2,000 miles to the west. All of this gadgetry is backed by a wraparound view of Chicago's skyscrapers, with Lake Michigan shimmering in the distance. In this technology showcase, the shopping cart looks out of place and a little forlorn. But it reminds Rayid Ghani and his small team of researchers of their key mission: to predict the behavior of people like my wife, and you, and me as we make our way through stores.

Ghani made a splash in 2002 with a study of how a clothing retailer like The Gap or Eddie Bauer could automatically build profiles of us from the things we buy. This sounds simple, but it adds a thick layer of complexity to data mining. If you unearth an old receipt gathering dust in your bedroom, you'll see that one afternoon a few months ago you bought, say, one pair of gray pants, two cotton shirts, and some socks. What can the retailer possibly learn about you from this data? That you're a human being with a body and, presumably, two feet? They take that much for granted. That you spend an average of $863 per year in the store? That's a tad more interesting. But if each one of the items you bought carried a bit more contextual information, what computer scientists call a layer of "semantic" detail, much more of you would pop into focus.

Let's say the pants are tagged as "urban youth." With this bit of knowledge, the system can move beyond your spending habits and start to delve into your personal tastes—much the way Amazon.com calculates the kind of reader you are from the books you buy. A clothing system with semantic smarts can send you coupons for garments that appeal to urban youth. It can track the proclivities of this "tribe" (that's a word marketers adore). And depending on the store's privacy policy, it might decide to sell that data to other companies eager to market songs or cars to the same group. Some, as we'll see later, might even use tribal data to push members toward one political candidate or another. Complications? No doubt. Maybe you're a 55-year-old woman who bought that pair of pants for your 16-year-old son. Maybe he hated them. That's not really you in the receipt, and it's not him either. Faced with such complexity and contradictions, machines need smart and patient teachers to guide them in making sense of us.

That's how Rayid Ghani views himself—as a personal tutor for the idiot savants we know as computers. Ghani is short, a bit round, and quick to smile. He's one of the friendliest tutors his students could hope for (not that they'd notice). A Pakistani who studied at the computer science powerhouse Carnegie Mellon, Ghani would seem to fit right in with the Numerati. But in their rarefied ranks, he's missing a standard ingredient: a doctorate. Having "only a master's" in his circle is viewed as a handicap. But the 29-year-old outsider has grown accustomed to clawing his way upward. The son of two college professors in Karachi, Pakistan, he applied to American colleges fully aware that he could afford only those offering a full scholarship. He landed at the University of the South, in Sewanee, Tennessee. Ghani calls it "a liberal arts college in the middle of nowhere." Hardly the ideal spot for a budding computer scientist, it is better known for its theology school. But one summer, Ghani won an internship at Carnegie Mellon, in Pittsburgh. He plunged into a world where classmates were teaching cars to drive by themselves and training computers to speak and read. He developed a passion for machine learning. Upon graduation from Sewanee, he proceeded to a master's program at CMU. Ghani was in a hurry. He started publishing papers nearly as soon as he arrived. And when he got his master's, he decided to look for a job "at places where they hire Ph.D.'s." He landed at Accenture, and now, at an age at which many of his classmates are just finishing their doctorate, he runs the analytics division from his perch in Chicago.

Ghani leads me out of his office and toward the shopping cart. For statistical modeling, he explains, grocery shopping is one of the first retail industries to conquer. This is because we buy food constantly. For many of us, the supermarket functions as a chilly, Muzak-blaring annex to our pantries. (I would bet that millions of suburban Americans spend more time in supermarkets than in their formal living room.) Our grocery shopping is so prodigious that just by studying one year of our receipts, researchers can detect all sorts of patterns—far more than they can learn from a year of records detailing our other, more sporadic purchases. (Most of us, for example, buy zero cars and zero TV sets in any given year.)

Three years ago, Ghani's team at Accenture began to work with a grocery chain. (They're not allowed to name it.) This project came with a windfall: two years of detailed customer records. The stores left out names, ages, and other demographic details, but none of that mattered. The 20,000 shoppers Ghani and his colleagues studied were simply numbers. But by their behavior in the stores, each number produced a detailed portrait of a shopper.

Let's assume you're one of those nameless shoppers. What can researchers learn about you? As it turns out, plenty. By the patterns of your purchases, and the amount you spend week after week, they can see if you're on a budget. They can calculate your spending limit. If they add some semantic tags to the data, they can draw other conclusions. When they see you starting to buy skim milk, or perhaps those miracle milk shakes, they can infer that you're on a diet. And they have no trouble seeing when you lapse. That carton of Ben & Jerry's in your cart, or the big wheel of Roquefort, is a giveaway. But wait! Maybe it's the holiday season, or your birthday. A few more weeks of receipts will spell out whether you're just cheating a little or in free fall. All of this they can do with the kind of statistical analysis an eighth grader could understand.

It gets a bit more complicated when they calculate your brand loyalty. Let's say you like Cherry Coke. You lug home a 12-pack every week. How much would Pepsi have to slash the price of its Wild Cherry Cola to entice you to switch? Ghani and two colleagues, Katharina Probst and Chad Cumby, watch how the shoppers respond to sales and promotional giveaways. They score each shopper on brand loyalty, and even loyalty to certain products within a brand. Some people, they've found, are loyal to certain foods, such as Kraft's macaroni and cheese. But does that loyalty extend to other Kraft products? For a certain group of shoppers, it does. The Accenture team takes note.

What they have on their hands is an enormous catalog of the eating habits of a small group of urban Americans in the first years of this century. Anthropologists of a certain bent would feast on it. But what good does it do a supermarket to know that you, for example, have a $95 weekly budget, are fiercely loyal to Cheetos, and flirted with the Atkins diet last barbecue season? What can they do with all that intelligence when they don't do business with you until you show up, loyalty card in hand, at the checkout counter? At that point, you've done your shopping. The chance to offer you promotions based on your profile has passed. Sure, they can throw a few coupons in your bag. Maybe you'll remember them on your next visit, but probably not. This is why, until now, supermarkets have virtually ignored the records of individual shoppers. They had little opportunity to put them to use.

The real breakthrough will come when retailers can spot you grabbing an empty cart and pushing it into the store. This has been a grocers' dream for decades. In a previous life in the 1990s, that sad little shopping cart at Accenture was a proud prototype of a "smart cart," one that allowed shoppers to swipe their loyalty cards through a computer attached to the cart, which would then lead them to bargains. "Everyone tried to do it," Ghani says. The attempts fell flat. The computers were too pricey, the analytics primitive. But computers are far cheaper now. Companies like Accenture are betting they can make systems so smart that shoppers will view the new smart cart as a personal assistant.

The first of such smart carts are just starting to roll. Stop & Shop is testing them in grocery stores in Massachusetts. Carts powered by a Microsoft program are taking their first turns in ShopRite supermarkets along the East Coast. The German chain Metro is launching them in Düsseldorf. And Samsung-Tesco, a Korean-British venture, has them operating in Seoul. A few things we know even at this early stage. For one, a computer on a shopping cart can ill afford to make dumb mistakes. This sounds axiomatic, but the fact is, we've long given grocery stores the benefit of the doubt when they offer us fliers and coupons that don't match our needs or wants, since they don't pretend to know them. But if a shopper has been buying skim milk for a year and the personalized cart insists on promoting half-and-half, the shopper may well view the smart cart as idiotic (and revert to the traditional dumb cart that specializes in rolling).

The other extreme? If these carts get too smart, we'll likely view them as creepy. I can just imagine rolling through my neighborhood Kings, when the cart starts flashing a message: STEVE: Hurry to aisle three for bargains on two of your favorite FUNGAL MEDICATIONS, plus this bonus SELECTION for the fungus you're most likely to contract NEXT! At that point, I'd be inclined to push it out to the street and under the wheels of an oncoming truck.

Setting aside such troubling scenarios, here's what shopping with one of these carts might feel like. You grab a cart on the way in and swipe your loyalty card. The welcome screen pops up with a shopping list. It's based on the patterns of your past purchases. Milk, eggs, zucchini, whatever. Smart systems might provide you with the quickest route to each item. Or perhaps they'll allow you to edit the list, to tell it, for example, never to promote cauliflower or salted peanuts again. This is simple stuff. But according to Accenture's studies, shoppers forget an average of 11 percent of the items they intend to buy. If stores can effectively remind us of what we want, it means fewer midnight runs to the convenience store for us and more sales for them.

Things get more interesting when store managers begin to manipulate our behavior. Rayid Ghani opens his laptop and shows me the supermarket control panel that he and his team have built. "Let's say you want four hundred shoppers to switch to a certain brand of frozen fish," he says. With a couple of clicks, the manager can see how many shoppers at the store buy this item. They sit in groupings known in marketing lingo as "buckets"—in this case, the frozen-fish bucket. Let's say it includes 5,000 shoppers. Among that group are those who buy rival brands of frozen fish. They're the target audience, and they sit in three smaller buckets, say, 1,000 shoppers per rival brand. Of those shoppers, one-third appear to be brand loyalists. It would likely take big discounts to pry them from the fish they usually buy. But the others, some 2,000, are more flexible when it comes to brands. They switch easily and often.

These buckets, as you can see, are getting increasingly refined. Now we're down to the brand-fickle buyers of certain types of frozen fish. Ghani plays at the controls. If he cuts the price by just 50 cents a pound—and sends word of the discount to their smart carts—he can entice a projected 150 of them to jump to the target brand. Ghani lowers the price by another 75 cents. At that level, an additional 300 bargain-hunters would line up to buy the fish. The manager can play with endless variables. He can adjust the formula to raise profits, to goose sales, to promote brands, to slash inventory. It's a virtual puppet show, all of it based on probability. The puppets, needless to say, are mathematical representations of us.

Let's say you're notoriously fickle when it comes to brands. Even the smallest fluctuations will push you from Cheerios to Wheaties and back again. If the manager is interested in slashing inventory, you're likely to be in the first bucket he picks up. You're an easy sell. But if the goal is to switch your allegiance from one brand to another, you're a lousy bet. No offense, but you're disloyal, at least in this context. You'll pocket the discount and abandon the brand the very next time you can save a dime. The manager might fare better promoting the discount to those who stick to brands a bit longer than you do. Naturally, they're in another bucket.

You may also lose out on discounts if you hew to a weekly budget. Let's say you spend about $120 a week on groceries. The system calculates that you're on a budget because, say, 87 percent of the time you spend between $113 and $125 a week. If you're not restricted to a formal limit, you might as well be. Assume that the manager is eager to get rid of a mountain of detergent moldering in the warehouse. He's offering jumbo boxes at two for the price of one. Should he send the word to your screen? Maybe not, Ghani says. The reason is simple. For every dollar you spend on discounted products, that's one less dollar you have in your budget to spend at full price. That hurts profits. To get rid of that detergent, it's smarter to target people in freer-spending buckets.

Among the most unpleasant buckets a manager must confront are those loaded with "barnacle" shoppers. That term comes from V. Kumar, a consultant and marketing professor at the University of Connecticut. Barnacles, from a retailer's perspective, are detestable creatures. We all know a few of them. They're the folks who drive from store to store, clipped coupons in hand, buying discounted goods—and practically nothing else. Kumar calls them barnacles because, like the mollusks clinging to a ship, they hitch free rides and contribute nothing of value. In fact, they cost the retailer money. With all the consumer data pouring in, Kumar says, it's becoming a snap to calculate a projected profit (or loss) for each customer. Kumar, who sells his advice to Ralph Lauren and Procter & Gamble, says that retailers should "fire" customers who look likely to drag down profits.

This doesn't mean hiring musclebound bouncers to block these shoppers at the door. But retailers can take steps in that direction. They can start by removing barnacles from their mailing lists. Increasingly, they'll also have the means to make

adjustments inside the store. If bona fide barnacles are pushing smart carts through a supermarket, for example, it might
make sense to fill their screens with off-putting promotions
for full-priced caviar and truffles. (Discouraging unwanted
shoppers is far easier on the Internet. Already, online merchants are assailing their barnacles with advertisements. And
if these bargain hunters click to browse the pages of a book or
gawk at the free photos on a paid-porn site, they get shunted
to the slowest servers, so that they wait and wait.)

  If you think about it, barnacles thrive in markets where
we're all treated alike. They feast on opportunities that the rest
of us, for one reason or another, miss. But now retailers are
gaining tools not only to spot barnacles but also to discriminate against them. Barnacles, of course, are the first to notice
when this happens. It's their nature to keep their eyes wide
open. And you can bet that they'll challenge this type of dis-
crimination in court. In a class-action suit in 2005, lawyers
representing some 6 million subscribers to Netflix, the film-
by-mail rental service, charged that the service was taking longer to send movies to its most active customers. Those were
the film buffs who paid a flat monthly fee of $17.99 for lim-
itless rentals and tried to see as many movies as they could
for their money. This involved watching a movie or two the
very day they arrived in the mail and rushing to mail them
back the next morning. (I know the routine; for my first few
months on Netflix, I was an eager barnacle.) Netflix officials
admitted that they favored less active (and more profitable)
customers with prompt mailings. And in a settlement, they
gave millions of subscribers a free month of service. But, significantly, they did not vow to change their barnacle-punish-
ing ways. They simply adjusted the wording in their rental
contracts.

Barnacles aren't the only creatures in Kumar's menagerie. He also warns retailers about "butterflies," customers who drop in at the store on occasion, spend good money, and then flit away, sometimes for months or years on end. They're unreliable, and retailers are warned to avoid lavishing attention on them. "You shouldn't chase the butterflies," the professor says. However, by studying their patterns of behavior, smart retailers may learn which butterflies they can turn into reliable customers—a bucket that Kumar calls "true friends."

As merchants learn more about us, it's going to be easier for them to figure out which customers to reward and which ones to punish. This won't make much difference to butterfly shoppers. They're oblivious. But in the age of the retailing Numerati, life for barnacles might get grim.

WITH ALL THIS TALK of butterflies and buckets, I ask Ghani, where is the individual? I expected to see myself modeled as a shopper, and here I am, sitting in buckets with other frozen-fish buyers and brand traitors. What's become of customization? Where's the fully formed mathematical model of the cheapskate who never pays the extra buck for yellow or red bell peppers? I'm talking about the reluctant clothes shopper, the one rushing through the mall with a tightening back who always takes two laps around the garage before finding his car? In short, where am I in all this data?

Ghani smiles as he delivers the bad news. There's no fully formed "me" in that data. There's no you, at least not yet. We exist in these databases as shards of our behavior, my hang-up with the bell peppers, your habit of casually tossing a bag of M&Ms onto the pile as you wait at the checkout. (By the way, those seemingly impulsive purchases, often accompanied by a

what-the-hell shrug, are no afterthoughts, Ghani's data shows.
Many shoppers buy the candy bars and breath mints more
predictably than they purchase milk or toilet paper.) In any
case, all of those pieces of our shopping selves reside in endless buckets with other people's slivers. Much as we might find
it flattering to sit in a unique bucket all by our lonesome, for
retailers there's no point. They don't have a customized marketing campaign for me or for you. They want to sell pork or
crew-neck sweaters. And for this, they'd like to bring together
1,000 or 50,000 people. Just because they like to microtarget
doesn't mean that they wouldn't rather reach lots of people
with the same message. They still love big numbers. They just
prefer to target customers more intelligently. It would be easy
to mistake these new buckets for the demographic groupings
marketers have worked with for decades: Hispanics, yuppies,
soccer moms, the super rich who inhabit the 90210 zip code
in Los Angeles. Those are buckets too. But there's a world of
difference.

In the old days, marketers knew next to nothing about
the individual, so they assumed that he or she shared values
and urges with similar people—those who also made six-figure salaries or had a last name with a vowel at the end. This
was a crude indicator. But given the information they had, it
was the best they could do. And in the decades of industrial-
ized consumption, in the 1950s and '60s, it wasn't half bad.
Choices were limited. Why bother learning about a person
if, chances were, he had little choice but to watch *The Hon-
eymooners,* eat one of three different kinds of peanut butter
on his sandwich, or buy a car that looked pretty much like a
Chevy? We have thousands more choices now, from the supermarket shelves to the remote on the TV, not to mention
the Internet. So marketers, as Tacoda's Dave Morgan demon-

strates, can shift their focus from who we are to how we behave. For this, they need the new buckets.

To see how different these new groupings are, consider the
demographics of these buckets we inhabit. Start by looking at
the skinflints who, like me, forgo the pleasures of red and yellow bell peppers. In this green-pepper bucket I'll wager that
I'm surrounded by people of all races. Both genders are represented (though I'd imagine, based on my family sample, that
more of us are guys). We drive all kinds of cars. Some of us
hunt; others would just as soon outlaw guns. The district at-
torney might be in there, sharing bucket space with the FBI's
most-wanted killer. You could say we have nothing in common, and you'd be absolutely right—except for one thing: our
behavior when it comes to buying bell peppers.

These bits of our behavior sit in thousands of buckets,
all of them created automatically by machines. Most of them
—like my green-pepper bucket—are never used. If you strung
all of your buckets one after the next, you'd see your own special combination, your unique shopping genome. Spend time
with microtargeting marketers these days, and you'll hear them
refer to these behavior patterns as a consumer's DNA. This
comparison is not fair or accurate, though it sounds tempt-
ingly simple. Unlike our genetic code, our behavior changes
all the time. We learn. (Who knows? After one tasty Moroc-
can meal I might be inspired to spring for a basket of exorbi-
tant red peppers imported from Holland.)

Still, forget those technicalities for a minute. Think of
buckets as genes. Each base pair of a gene (which provides instructions to produce amino acids) is described by combinations of two of four chemicals known as nucleotides. They're
represented by the letters A, G, T, and C. That basic code is
pretty simple. But there are key variations, both in the DNA

code for individual genes and in the 3.2 billion base pairs in the genome. To a large degree, those differences shape our bodies and our lives, distinguishing us not only from other plants and animals, but also from each other.

Since the 1990s, thousands of the world's leading mathematicians and computer scientists have been drawing up algorithms to comb through vast databases of DNA and other health data. They're looking for patterns in those billions of base pairs that might point to a proclivity for leukemia, creative genius, alcoholism, or perhaps a deadly allergy to peanuts. The research is still at an early stage, but scientists have built an enormous mathematical toolbox for linking symptoms to variations in the four building blocks of DNA.

Why does that matter to a grocer? For now, it doesn't. But let's say that a supermarket, a few years down the road, organizes each aspect of our shopping data into four groups. For example, we buy candy at the checkout

1. More than 90 percent of the time
2. From 25 to 89 percent
3. From 1 to 24 percent
4. Never

With modern computing, it wouldn't be that hard to organize thousands, or even millions, of our grocery-shopping habits into similar groups of four. They'll be arbitrary, much like the census or the categories on insurance forms. The point here, however, isn't to model one entire person accurately but instead to decode the patterns of human behavior. Consider the people who buy luxury chocolates. Is there anything in their purchasing behavior that appears to trigger chocolate lust? Grocers have wrestled with these questions for centuries. They make sensible correlations. Chocolate lovers might be interested in almonds. Catch them at the holidays and before Valentine's Day. But how about the correlations that humans wouldn't think to look for, such as the romance-movie lovers who clicked on Alamo car rental ads? How do grocers unearth those hidden links?

This is where the data-mining algorithms could come in and lead to randomized experiments with shoppers, Ghani says. Once the retailers have our behaviors grouped into four variables, they can retool one of these genomic algorithms and feed our shopping data to it. The computers whir through our purchases, looking at literally billions of combinations. The great majority are utterly senseless. Do people who buy both Brussels sprouts and sugared cereal also buy Swiss chocolates more than the mean? No sane person would bother looking for such a connection. That's why it's the perfect job for computers. Set them on a hunt, and they might find correlations we humans would never think to consider. Just as they've helped medical researchers find genetic markers pointing to certain types of breast cancer and Huntington's disease, they might tell grocers what kinds of fruit to promote to buyers of canned food or what types of magazines dog-food buyers tend to read. These suggestions may sound frivolous. But if a retailer can tweak promotions, bucket by bucket, and gain a boost of even 2 percent of sales, it's cause to rush down aisle seven and pop a magnum of Mumm's. They measure profit margins in this industry by the tenth of a percent.

As Ghani talks about shopping patterns and genomic researchers, I think about putting all the people we've been talking about—the grocers, the microtargeting advertisers, the mathematical geneticists—into one room. They wouldn't seem to have much in common. Yet they do. In nearly every industry, the data we produce is represented by ones and zeros.

It all travels through the same networks and vies for space in the same computers. This means that the mathematical tools used to analyze this data can cross disciplines and industries, from the barnyard to the aisles of Saks, almost effortlessly. This has a nearly miraculous multiplier effect—the brains working in one industry can power breakthroughs in many others. Researchers long isolated in different fields, different departments on campus, different industries are now solving the same problems. The analysis of networks, for example, extends from physics to sociology. In a sense, all of these scientists are working in one global networked laboratory.

All of which is to say that researchers whose tools will one day decipher the secrets of your shopping—perhaps the subconscious patterns you don't even know about yet—may not be working for Wal-Mart or Google or Ghani's team at Accenture. Today they might be studying earthworms or nanotechnology, or maybe the behavior of Democratic voters in swing states. For example, one researcher at Microsoft, David Heckerman, was hard at work building a program to comb e-mail traffic and identify spam. He knew that spammers systematically altered their mailings to break through ever more sophisticated defenses. He was dealing with a phenomenon similar in nature to biological mutations. His system had to anticipate these variations. Heckerman, a physician as well as a computer scientist, knew that if his tool could detect mutations in spam, it might also work in medicine. Sure enough, in 2003, he shifted his focus to HIV, the virus that causes AIDS. His tools, with their legacy in spam, could eventually lead to an AIDS vaccine. "It's the very same [software] code," he says. In the Numerati's world, breakthroughs can come from any direction.

. . .

CONSIDER FOR a moment the clothes you put on this morning. If Rayid Ghani and his colleagues had a picture of you as you made your way down to breakfast or out the door, would they know from your clothing what tribe to put you in? Chances are, they could come pretty close. Humans have specialized in tribal recognition since we climbed down from trees. It's a survival skill.

But how does Ghani teach that skill to a machine? Computers, after all, have to figure out what kind of clothing we're buying if they're going to classify us as dweebs, business drones, hip-hoppers, earth mothers, or whatever other fashion buckets the marketers create. It is true, of course, that armies of people could flip through these garments, giving each one a tribal tag. But this procedure would cost a bundle, and the workers (who themselves come from different tribes) would surely disagree on what's sexy, fashion-forward, or retro. Humans are just too subjective. This is a job for computers. However, when it comes to classifying clothing, Ghani says, machines fare no better than the most clueless of humans—at least for now. So the Accenture team in the Chicago lab has to cheat.

Here's how. They hire a group of people to teach the computer. These trainers slog through a questionnaire from an online department store catalog. For several hundred garments, they answer a series of multiple-choice questions. Is it formal or casual? Is it business attire? On a scale of one to ten, how sporty is it? How trendy is it? What age group is it for? On and on. Several people evaluate each item. This smooths over their individual quirks and produces a consensus. As the humans answer these questions, the computer learns about each piece of clothing. If it were human, perhaps it would be able to develop an eye for what's sporty and what's hip, and then be able to classify the rest of the fashion universe by itself. But computers don't yet have such discerning eyes. Instead, the machine focuses on the promotional language that accompanies each picture. *Zesty! Hot! Spring fever!* It learns to associate those words with the values spelled out by its human trainers.

In the end, the computer builds up a matrix of words, all of them defined by their statistical relationships to each category of clothing. *Bra,* to cite an obvious example, would have a near-zero probability of belonging in men's wear. In every example marked by the humans, it shows up for women. But that doesn't tell the computer whether a certain bra is sporty, casual, or Gen Y. For that, it must find clues in other words.

Ghani shows me the vocabulary his system has mastered. He calls up "conservative" words. The computer spits out *trouser, classic, blazer, Ralph,* and *Lauren.* Words that rank low on the conservative scale? Ghani calls it up and laughs. "*Leopard*! That's a good one." Others are *rose, chemise, straps, flirty, spray, silk,* and *platform.* I'd say the computer has figured out a thing or two. When Ghani asks it for "high brand appeal," *DKNY* and *imported* show up, along with that now familiar duo of *Ralph* and *Lauren.* (This system, Ghani explains, has no fancy understanding of context. Unlike other artificial intelligence programs, it is unburdened by grammar. It just plows through the English words it has encountered and pegs each one to a set of probabilities.)

Figuring out that a certain white blouse is business attire for a female baby boomer is merely step one for the computer. The more important task is to build a profile of the shopper who buys that blouse. Let's say it's my wife. She goes to Macy's and buys four or five items for herself. Underwear, pants, a couple of blouses, maybe a belt. All of the items fit that boomer profile. She's coming into focus. Then, on the way out she remembers to buy a birthday present for our 16-year-old niece. Last time we saw her, this girl was wearing black clothing with a lot of writing on it, most of it angry. She told us she was a goth. So my wife goes into an "alternative" section and — what the hell? — picks up one of those dog collars bristling with sharp spikes.

How does Ghani's system interpret this surprising deviation? Jaime Carbonell, a professor of machine learning at Carnegie Mellon, thinks about these issues a lot. In the early days, he says, consumers were often averaged. He noticed that Amazon.com, for example, saw that he was interested in Civil War history and in computational biology. So it combined them. He got recommendations for the history of biology and the north-south divide on some scientific question. "The average modeling doesn't work well," he says. "We're not the average of our interests." The newer approach is to use clustering software. This divides his interests into different groups and gives him recommendations based on each one.

Let's say my wife's purchases were clustered. The system could look at most of her purchases and conclude that she's a female boomer. The dog collar? It's what statisticians call an outlier. In these early days, it's something that's safer to ignore. But as analysis gets more sophisticated, it will latch onto those bits of our lives that appear to be deviations. After all, which details are more likely to lay us bare, our day-to-day behavior that appears "normal" or the apparent quirks that we often work to hide? A detective will opt for the outlier in a New York second. The marketer might too. But it's tough to make sense of such data with automatic systems.

In any case, suppose that next week my wife returns to the same store and buys piercing tools and green hair dye. At that point, the software might turn the spike collar she purchased, that apparent outlier, into its own cluster. So what would that new cluster tell us about her? Hard to say. Is she a middle-aged professional who commutes Monday through Friday in sober attire and then, on weekends, straps on the spiked collar and goes goth? Could be. Or perhaps she's buying for two people. Ghani says that some systems in grocery stores look at the different clusters and try to come to conclusions about the composition of a family. Others look at the different signals as varying dimensions of one person. Sometimes, though, "mutually exclusive" purchases in the same cart—small socks and big shoes—indicate that more than one person is involved.

Accenture's automatic fashion maven isn't yet grappling with such subtle distinctions. It's still in the research phase. But once this type of technology is in the marketplace, stores will have strong signals as to what types of shoppers we are. At the same time, they'll be compiling ever more detailed and valuable customer lists. As we'll see, plenty of other marketers, such as those in dating services or political groups, would pay richly for, say, a list of 10,000 trendy Gen Y women in Seattle, Chicago, or Miami. And yes, there will be lively markets, no doubt, for assorted varieties of goths.

LET'S SAY YOU go to a department store with a shopping list. If you come back missing a couple of items, the store has failed an important test. Even if you locate and buy everything on the list, your visit, from the store's perspective, falls short of an unqualified success. No, they want you to stumble upon countless temptations as you make your way up and down the aisles. In their dreams, you teeter up to the checkout under such a pile of serendipitous finds that you have to pay a young assistant or two to help you lug it to the car.

How to make that happen? The first step is to map our

migrations through the store. In the old days, some store managers and museum curators would gauge foot traffic by the wear on the floor tiles. Then they would redeploy their offerings to draw customers off the beaten paths. But that approach is a tad slow for the Numerati.

Ghani and his team have another idea. As we walk around the Accenture office, cameras hanging from the ceiling are tracking our every move. There are about 40 of them, Ghani says matter-of-factly. From my perspective, it's insidious workplace surveillance. With this kind of spy network installed in my skyscraper offices in New York, I think I'd find myself rationing my trips to the bathroom. But Ghani and his colleagues view the cameras as just one more experiment, this one to track workers and customers. The Accenture workers are offering themselves as specimens, and they don't seem to mind a bit.

This type of monitoring system isn't that relevant to Accenture's lab setting, where the flow of information counts for more than bodily movements. But Ghani sees growing numbers of cameras tracking the movements of customers and employees in big stores, hotels, and casinos. They could also find a home in factories. Such cameras are already installed as a security measure, Ghani says. So now it's just a matter of giving the camera another job.

With this type of snooping technology, managers can start scrutinizing our movements. In these early days, they focus more on overall patterns of traffic than on individuals. That's because today's cameras have foggy vision. They see us as little more than moving blurs, Ghani says. They'd be hard-pressed to identify our faces, even if we stood perfectly still, gazed up at them, and mouthed, ever so slowly, our names. Most automatic surveillance systems, which seem to recognize faces so well in the movies, don't yet work such magic in the real world. Douglas Arnold, director at the Institute for Mathematics and Its Applications at the University of Minnesota, says that facial recognition was oversold as far back as the 1960s. Researchers are making progress, Arnold says, but "if people start relying on facial recognition systems today, they're going to be burned."

So how will Accenture's cameras pick out individual workers and shoppers? Ghani introduces me to what he calls massive redundancy. This involves getting lots of cameras to work together as a team. Each one provides a bit of detail. It's a little like a group of witnesses who see a thief dash by. One might remember his red hat, another the bandage on his hand. A third points to the alley he ran down. In Accenture's case, the system can stitch together these smidgens and come up with a guess as to who each blur is most likely to be. They can be pretty sure, for example, about the identity of a short, dark-haired figure wearing a blue shirt who emerges from Ghani's office accompanied by a taller stranger with an oddly stiff neck (me, bad hotel pillows). Stature, colors, and the patterns of movement all indicate that the person is Ghani. The system makes similar calculations about the other Accenture employees on the floor. This produces truckloads of visual data. Accenture's computers use that information to feed all sorts of analysis. They can create charts showing each person's migratory patterns, social hubs, and yes, even bathroom visits. Similar analysis could be focused on us as customers. In time, perhaps a store will recognize us by our movements in the aisles as likely butterflies or barnacles, or even potential shoplifters. And as the facial recognition systems improve, they may spot the barnacles among us the moment we enter the store.

If cameras don't pick us up, a radio technology known as

RFID just might. These are little computer chips fastened to a piece of merchandise, a shopping cart, or even a customer loyalty card. Each chip has a unique number, identifying the item or the shopper. But unlike a bar code, which has to be passed under a scanner, these chips can be read by radio signals sent by an automatic reader in the area. It's great for logistics. Open a huge cargo truck, and instead of piling through it and scanning each bar code, the chips all transmit their data at once. The detailed contents of the shipment pop up in a split second.

These same chips can track us in stores and at conventions. AllianceTech, a company in Austin, Texas, puts these radio tags into the ID badges people wear around their necks or clipped to their lapels at trade shows. The company also puts receivers into the booths they visit. Then, when IBM or Texas Instruments wishes to know who visited its booth, AllianceTech can give them the names of the people (at least those who agreed to participate), their companies and industries, and the amount of time they spent in the booth. They can even see how much time these people spent visiting a rival's booth. If you look at the flow of customer data, it's as if the whole trade show is taking place on the Internet.

Imagine what would happen if retail stores used the same technology. Some are moving in that direction. Germany's Metro, the world's fifth-largest retailer, is equipping smart carts with radio transmitters in several of its stores. The technology, says Albrecht von Truchsess, a spokesman for Metro in Düsseldorf, is meant to provide shoppers with enhanced service but not to compile their shopping-related data or to profile them. (Data privacy is a far more explosive issue in Europe than in the United States.) As shoppers push the smart cart, they scan the bar code of each item they toss in. This information is sent by wireless connection to the computer, and, much like a driver cruising through an automatic toll-booth, the shopper can roll the cart out of the store without stopping to pay. The technology has taken care of that.

By mapping the trail of those scans, Metro can trace the minute-by-minute wanderings of each shopper. Even without building personal profiles, Metro's analysts will be able to study patterns. They may discover that many of the most carefree, sky's-the-limit shoppers never encounter the display in aisle three promoting sinfully rich (and expensive) Belgian chocolate. Just like a website, the store has plenty of options to entice the consumer: It can flag the chocolates on the smart cart screens. Or it can tweak the layout of the store, placing the chocolates along the pathways most popular among spendthrifts. Pity the dieters who dare to shop in the markets of the Numerati.

CHAPTER 3

Voter

QUICK. Who did you vote for in 1996?

Does that question inspire just a trace of panic? Do you worry that you might not remember, or perhaps that you'll blurt the name of a candidate who wasn't even running? Maybe you remember perfectly well, but you're afraid that if you tell me, I will hound you with aggressive follow-up questions. You voted for HIM? What in the world was going through your mind?

If you experience these fears and anxieties, welcome to the club. You're in the majority. Those who actually enjoy politics, who naturally think about the world the way politicians do, are a minority in just about every country. But because members of this cozy club run the political realm, they tend to analyze politics as if everyone else viewed it with equal fascination and zeal, and the same focus on issues. That's why they fail to connect with voters, says Joshua Gotbaum, or even to understand them.

Josh Gotbaum is a card-carrying member of that minority. He served, as a young man, in the Carter administration, and he returned in middle age to work for President Clinton. He held senior posts at the Pentagon, the Treasury Department,

and the Office of Management and Budget. When the Republicans run the show, he makes money and works for charity.
At one time, he was a partner at Lazard Frères, the investment
bank. After the terror attacks of 2001, he headed the September 11 Fund. Later, he moved to Honolulu and led the effort
to pull Hawaiian Airlines out of bankruptcy. Now, as I talk
to him, he's attempting to resuscitate an education start-up in
New York, from offices high above Wall Street. But he's eager
to get back to working in government. Naturally, the key is to
find more people in the U.S. population who could conceivably vote for a Democratic candidate and to come up with just
the right pitch for each voter. The trouble, he believes, is that
millions of potential Democrats are camouflaged. For one reason or another, they pass for Republican voters. Some live in
McMansions with fairway views and drive Hummers. Some
no doubt carry weapons, revere the military, or spend much of
their free time praying. Others are staying hidden because they
haven't been much impressed to date by Democratic candidates. To turn these people into Democratic voters, Gotbaum
says, his party needs to surpass the Republicans in pinpointing
potential supporters from within massive databases.

The Republicans set the standard for political microtargeting in the 2004 elections. First, they framed the issues in
a fresh way, avoiding much of the policy verbiage that bores
or bothers most voters to distraction. They focused on simple
desires closer to the heart than the head. Things like feeling
safe, loving their country, and being surrounded by people of
faith. They spent millions on polling, and they used what they
learned to unearth their target voters. Matthew Dowd, an aide
to President Bush's longtime operative Karl Rove, joined with
two coauthors to detail this triumph in the 2006 book *Applebee's America.*

Now Gotbaum, his auburn hair and mustache tinged with

gray, has set up his own political firm, Spotlight Analysis, and raised $1.5 million in venture funding. He views it as the Democrats' riposte. Success with Spotlight could be his ticket back to power. The way he sees it, the party that figures out how to harness the power of data, and the brains of the Numerati, is going to win at the polls. It's not that microtargeting will corral the 50 million or 60 million voters that a presidential candidate needs to win. The traditional approach—blizzards of TV ads and immense phone and door-knocking campaigns—bring in the lion's share. But in races decided by one or two percentage points, or less, the party that pinpoints a few thousand individual voters here and there could come out on top. If the data we emit gives off even the slightest whiff of "swing voter," the political Numerati will be hot on our tracks.

This has implications for the "decideds" too, as politicians start using their data to deploy limited resources more efficiently. If you come across as a safe vote, the candidates you endorse can direct their promises and stump speeches toward wavering supporters. This has long been the case. But traditionally, politicos have appealed to entire fence-sitting clans, crafting one appeal for retirees in Florida, for example, and another for worried autoworkers in Michigan. Now, though, the Numerati are collecting far more data on these swing voters and can study them with greater precision. This means they can place them into ever-smaller groups. As they create and test messages aimed at each sliver of the electorate, the science of the Numerati is supplanting the folk wisdom of the precinct chiefs.

THIS IS THE CHALLENGE Gotbaum faces as he tries to find votes: how do you calculate what moves people politically if, for many of them, the entire subject is toxic? Compared to politics, shopping's a cinch. People long for things. They pick them up. They buy them. They leave clear records. When Rayid Ghani and his team at Accenture dissect your supermarket receipt, they can see that every two weeks you buy, say, a bag of bright green Granny Smith apples. In a sense, you vote for those apples with your credit card. Based on the patterns of other apple buyers, it doesn't take a big leap on their part to predict other foods you might like. They're working well within the realm of the data you provide. The comparison is what we might call "apples to apples."

Now consider politics. Many people don't like to think or talk about it. They switch the channel or turn the page. With the explosion of new media, they have literally thousands of news and entertainment choices—many of them, in their opinion, a lot more fun than politics. What's more, U.S. citizens resist formal efforts to measure their political sentiments. If you think I'm kidding, consider this. Pollsters say that as few as 12 percent of Americans bother to respond to their phone calls. What's worse, from a data hound's perspective, those moved to leave their house on a rainy day in November vote behind a curtain. It's secret.

This means that political operatives must dig into other data to find potential voters. Traditionally, they've settled for proxies for our behavior. Our neighborhoods used to be a good predictor of voting behavior. The same with race. But most of those big categories are breaking down. We walk less and less in lockstep, and we have more choices. For the kind of analysis Gotbaum wants, he must delve deeper. He has to find out not just where we live and work but what we love, what we fear, what we feel deep down about squishy subjects like community and country. The mathematics of politics, strange as it might sound, has to extend beyond the grasping fingers of our inner shopper and plunge into terrain that lies closer to the soul. To get there, researchers must move far past mouse clicks and Google queries, those commodity piles of personal data that course through the Internet. This involves asking lots of questions.

Here's one, just to give you an idea. What is your community? Conjure it up in your mind. If you could draw a picture of it, what would it look like? Would it feature people waving from windows and porches in the group of houses up and down your street or the apartments on your block? Are these physical neighbors your community? Or is it a group with shared values, perhaps based in a church? Maybe your community is a widespread Internet group gathered around a blog that focuses on Dostoevsky's novels or Chianti. Maybe you have an even more expansive view of community, which includes all of us who live and die on this blue lump of rock circling the sun. For some, our community extends to animals. Yet just imagine the blank looks a politician will receive in certain quarters if she evokes a community that includes pigeons and pilot whales. Some voters will write her off as an alien. Yet for others, she's a kindred soul. It's second nature for skilled politicians to tweak their message, or even their accent, for different audiences. But how can you read the crowds when they're dispersed across networks?

Think of the hazy words and concepts that pop up in politics. Freedom. Democracy. Justice. Security. Opportunity. Human rights. Wealth. Politicians reflexively drop them into stump speeches and TV ads. Yet they all spark wildly different responses. For some, the word *justice* means executing murderers. For others, it's giving poor children equal opportunities in school. If politicians could group people according

to how they view these concepts, they would better understand the role that each person sees for government in society.
That's what we express when we vote. And with that knowledge, politicians could craft messages that speak to our values
and appeal to our concerns.

This was a simpler job for past generations of politicos,
because we Americans organized ourselves in clear groups
and worked hard to assimilate. Take my parents. In 1954, they
moved to a leafy suburb of Philadelphia with their three girls.
(I wasn't around yet.) They promptly went about the business
of fitting in. They bought a big red Plymouth station wagon
and put their names on the waiting lists for two of the snooti-
est clubs on the eastern seaboard, one for tennis, the other for
golf. They began attending a venerable Episcopalian church
favored by the established order. They subscribed to the (now
defunct) Republican paper, the *Evening Bulletin*. And like
most of their new neighbors, they registered locally as Republicans.

If one of the political Numerati at the time (such as they
were) had been told to create a mathematical model of my
parents, he probably would have asked, "Why bother?" There
was little need to customize anything for them. They were doing all the work themselves by adjusting to the local norms.
They were squeezing into an off-the-shelf algorithm. It defined a large and politically moderate group known as Rockefeller Republicans. Many of these families had been Republican since the days of Abraham Lincoln, when Republicans in
the North led the fight to defeat slavery and save the Union.
From oil paintings on the walls of our living room, a couple
of our nineteenth-century ancestors stared down sternly at us.
I always assumed they were Republicans. Wasn't that what we
were?

In the early 1960s, my father ventured into local politics and won a spot on the town board. He could tell you, block by block, which houses could be counted on for Republican votes—and which doors to skip on election day. (Professors at nearby colleges tended to vote the other way.) Things were pretty simple. But with the growth of the civil rights movement and the Vietnam War, changes started to percolate in our Republican household. The party's nominee for president in 1964, Senator Barry Goldwater, stood far to the right of my parents, especially on civil rights. My mother feared he'd push us into nuclear war. So she became a "Republican for [Lyndon] Johnson," and she broadcast her support for the Democrat by placing a bumper sticker on our station wagon. My father, hounded by a local conservative columnist, lost his next election. Over the next few years, it was as if a 100-year-old dam had broken. My parents took buses to Washington to march against the war. They quit the country clubs and the big church. And when they moved out of the house in the early 1970s, they went to considerable trouble to find African American buyers, the first in the neighborhood. At some point during this process they became Democrats.

Look at them from a political pollster's point of view, though, and it was hard to tell they weren't still Republicans. They lived in privileged neighborhoods and rode the stock market boom of the 1980s. My father continued to climb aboard the same commuter trains that traditionally shuttled suburban Republicans back and forth from Center City. How was a statistician to know that instead of traveling to Bar Harbor or St. Moritz, this particular couple was vacationing in Nicaragua, joining a "human shield" to defend the Marxist government there from the American-backed Contra rebels? They were camouflaged. Yet even though my parents had abandoned the Republicans, the Democrats didn't know exactly what to make of them either. Like many Americans (of the minority who are politically engaged), they had broken loose. Politics no longer described their inherited identity. It now fit into a long menu of consumer options. Political mailings arrived every day, along with the credit card come-ons and gardening catalogs. They sifted through them, looking for candidates and issues that appealed to them, and threw the rest into the trash.

If you think about it, this free-range shopping has moved into almost every facet of our lives. People now shop for neighborhoods, religions, and cuisines that suit their lifestyle. Should you keep preparing the chimichangas or goulash your grandmother made or become a vegan? It's a choice. Millions of us now shop for climates and even countries, settling in Vancouver or Barcelona, or retirement havens on the lakes outside Guadalajara. Even the slope of our nose has become a choice. So why not pick and choose in politics too? That's what more and more of us do. This means that politicians, who used to locate us by our old groupings, must now find the new tribes and communities we form, often ones based on interests or values. The words *Democrat* and *Republican* are now foggy old terms that fail to describe most of us. As far as politicians are concerned, millions of us are lost (or don't care). Like other marketers, they're attempting to track us down by following our data. Only then can they engage us.

This is a tall order. How will Gotbaum, with his modest funding, dig down deep within you and me and masses of other voters, and scope out our philosophical moorings? Even if he talks to 1,000 people, or 10,000, how can he extend what he learns to chart the political currents of an entire nation? And if most of us don't feel like thinking or talking about politics, what kind of data will express our unspoken (and often unknown) political convictions?

Gotbaum smiles and takes me back to late 2005. "It was pretty clear," he says, "that Republicans had been spending money to understand the independent voters. I got back from Hawaii, and I thought, 'Corporations spend a zillion dollars on market research. Why can't that research be made available to Democrats?' That was my notion." So early in 2006 he called in two market researchers and a pair of political pollsters, and he dispatched them to survey the values of Americans.

The raw material for this type of research—the details that make up our lives—is collected and sold by a fast-growing group of data companies. ChoicePoint, headquartered in the Atlanta suburb of Alpharetta, Georgia, is a prime example. The company quietly amasses court rulings, tax and real estate transactions, birth and death notices. Many of these records have existed for centuries in file cabinets and courthouse ledgers. But ChoicePoint employs an army of data collectors to harvest those facts, sometimes writing them down by hand. Then they put them into digital files.

Files that used to exist on different pieces of paper in different buildings can now be brought together. Our profiles start to take shape, and they can glide around the world on networks. Human resources managers dig up our Choice-Point files to see if we're lying on our résumés or have neglected to mention that year of hard time on Rikers Island. While ChoicePoint sticks to identity data, plenty of other data companies, as Robert O'Harrow Jr. writes in his authoritative book *No Place to Hide,* add flesh and blood to those bare bones. One of the largest, Acxiom of Conway, Arkansas, keeps shopping and lifestyle data on some 200 million Americans—nearly every adult in the country. Acxiom knows how much we paid for our house, what magazines we subscribe to, and which book we bought two days before taking that trip to Club Med in the Alps. The company buys just about every bit of data about us that is sold, and then they sell selections of it to anyone out to target us in a campaign.

Those companies provided data, but Gotbaum needed someone to turn it into something useful—a tool to find pockets of swing voters. Enter J. Walker Smith, president of Yankelovich. Since the early 1970s, this research company, based in Chapel Hill, North Carolina, has been surveying the shifting attitudes of the American public. It compiles these trends in a report called *Monitor,* whose motto says it all: "What in the world are they thinking?" It was this type of research, Gotbaum believed, that would enable him to create a new type of political profile—to locate millions of potential Democratic voters.

At the heart of *Monitor* are three sets of questions, Smith tells me, from his offices in Atlanta. First, what does the future hold? Does it look promising? Scary? Exciting? Bleak? The second question centers on how people will fare in that future. How do they define success? A healthy family life? Making piles of money? Grappling ever upward in the business world? Gaining respect as a good neighbor, an upstanding member of the community? The third question focuses on the skills and assets people believe they will need to achieve success in the coming years and decades. These questions, vague as they sound, get to most of the issues that have us gnawing on our fingernails by day and tossing and turning at night: What's ahead, both good and bad? Where do we want our lives to take us? How can we make this happen? Marketers in the auto and travel industries find these studies useful in deciding which features to promote in their products. If the public mood in the *Monitor* takes a skittish turn, for example, a carmaker might emphasize antilock brakes and a chassis that can withstand a bruising encounter with an oak tree.

"We suggested to Josh and his team at Spotlight that these values could be used to profile voters," Smith says. The idea was not just to identify voters in a key state or congressional district but to group every single voter in the country according to a set of values. Gotbaum was on board from the start. But others on the team feared that a consumer survey full of vague "values" questions would lead to a political dead end. Smith himself couldn't guarantee it would work. "We wouldn't know until we tried it," he says.

So they went ahead. In a marathon session in early 2006, they drew up a survey that mixed the traditional Yankelovich battery with lots of other questions. They asked, for example, if "taking care of our country's children" should be the number one priority, if the country should do "whatever it takes" to protect the environment, if we need to "rebuild and strengthen our shared vision" on what it means to be the American community. They also planned to ask voters which party they had voted for in recent elections. In the end, they figured it would take an average voter about 35 minutes to answer the 140 questions on the survey. This was a lot to ask, but Gotbaum trusted that even people who hang up on political pollsters and other phone spammers would take time to respond to questions about their personal values. "People like to talk about themselves," he says.

The next step was to choose 3,000 people to participate in the survey. These would become the test group, the gold standard. Interviewing these 3,000 would be the painstaking work of humans. Computers would later build upon the patterns culled from those 3,000 to classify, at warp speed, the rest of us. Naturally, this group had to reflect a cross section of voters. Gotbaum made sure that each one of them was already represented on Yankelovich's database of American consumers. That way, his team would have fat dossiers on every person surveyed. They would be able to slice and dice the group every which way. If those most concerned about taking care of the country's children appeared different in any way—if they lived in certain types of neighborhoods, went to church more often than others, even if they drove Pontiacs or ate sushi in disproportionate numbers—Gotbaum wanted to know about it. There would be no blank slates in this sample. His pollsters would start with telephone interviews and then reach as many of those people as possible for follow-up questions via the Internet. (This idea also generated heated arguments, Gotbaum says, because it's hard to find a representative population on the Net. Those responding to Internet surveys are likely to be slightly richer, younger, and better educated than the voting population as a whole. But small start-ups like Spotlight take such shortcuts to save money and later adjust the results for the bias.)

By summertime, the surveyors had conducted their interviews. The Spotlight team was sitting on reams of new data. Gotbaum asked Yankelovich and one more corporate research firm, StrataMark, to analyze the answers about values. "I told them to ignore politics," he says. "Just tell us, if you were going to segment this population on attitudes and values, what would the segments look like?"

Both researchers returned a month later with parallel results. Each of them concluded that voters fell into six clear values categories. They tossed out the smallest one, the 29 voters who appeared to be either angry or alienated from the political process. No sense in building campaigns for them. That left five groups. As Smith describes them, they were focused around these values:

1. Extending opportunity to others
2. Working within a community
3. Achieving independence
4. Focusing on family
5. Defending righteousness

It doesn't take an advanced degree in political science to guess which way a couple of these groups would vote. "Extending opportunity" sounds like a code word for bleeding-heart Democrat. And that righteous fifth group appears unmistakably conservative.

Gotbaum saw as much himself. He hardly bothered with the extremes. They were firmly in the Democratic and Republican camps. Interestingly, the core Democratic group, with 37 percent of the total, was more than twice the size of its Republican counterpart, which only had 16 percent. This meant that Republicans were managing to attract many more of the middle segments, the swing voters. The three swing sectors, grouped around community, independence, and family, accounted for 47 percent of the electorate. These were the voters who guided the country from right to left and back again, from Reagan to Clinton to Bush, to . . . well, whichever candidate figured out how to connect with them. Some of these people saw government as a potential ally; others viewed it with suspicion. They differed greatly on religion. And yet lots of voters in each of these groups were clearly open to persuasion. What moved them? To win them over, Gotbaum believed, Democrats had to appeal to their core values: community, independence, and family. If he and his team could just pinpoint thousands of these voters in a state or congressional race, the Democrats could craft custom-made messages for each group. The way Gotbaum planned it, the Spotlight system would locate perhaps 20,000 or 30,000 members of a certain value group in a tight congressional race. If research showed that many in this group listened to a certain gospel radio station or perhaps watched a cooking show on cable TV, the campaign would reach them with a targeted ad. A more precise approach would be to bombard each of the voters with carefully calibrated letters and brochures. Gotbaum didn't yet know exactly what those ads would focus on. For that, he needed to slice the data more thinly.

The Spotlight team studied each voter's views to see how deeply they cared about these issues of politics and values. Some in each group were heavily committed, hard-core, while others appeared to focus more on other aspects of life. Still working with their original test group, Spotlight split each of the five segments in two, one group that was more committed, one that was less. The more committed voters might be harder to swing. But they were also more likely to pay attention, and to vote. Moreover, the commitment, especially in the middle groups, didn't necessarily signal unswerving loyalty to one party or the other but instead to particular values. Following the traditions of consumer marketers, Spotlight gave each of these ten groups a descriptive name, such as "Barn Raisers," "Hearth Keepers," or "Inner Compass." As Gotbaum tells it, these are tribes. All of us are members of these values tribes (and countless others) without even knowing it. These tribes of ours have no logos, no history, no home turf, no particular cuisine or religious leaning. They span all races and ethnic groups. In that sense, they're a bit like those buckets of broccoli eaters or Mars Bars buyers in the super-markets.

I stop and try to think of anyone I know who might be a "Hearth Keeper." Such people, according to the Spotlight blurb, focus on family and faith. But they resent attempts to politicize these values. They're the less committed of the independence segment, not quite as rambunctious as the Barn Raisers. Looking at a color-coded chart, I see that Barn Raisers tend to be more entrepreneurial and mingle more within their communities, while Hearth Keepers, as their name would indicate, focus more on family satisfaction. Both groups adhere to "faith-based living" and lean Republican. But Hearth Keepers are more likely to keep it to themselves, and they resist "marketing intrusions into their private lives." (They don't sound like the most promising group for one of Gotbaum's microtargeted campaign calls. "I know you don't like this kind of call," the volunteer's prompt might read. "But this one is different . . .")

I look around the coffeehouse where I'm writing. Any Hearth Keepers sipping cappuccinos around here? The college student to my right — shaggy hair, flannel shirt, anthro books on the table — has his feet stretched out on a chair. He's wearing black low-cut Keds. If this is a Hearth Keeper, he's hiding it. Next to him, a middle-aged man with horn-rimmed glasses is just getting settled. He wears a neat V-neck blue sweater vest over a freshly ironed powder-blue shirt. Could be, I think. It's Sunday, he may have come straight from church . . . But shouldn't he be at home, with his family? Now he opens an Apple laptop (exactly like mine), and I'm quickly revising my judgment. I could be in the presence of a techno-friendly Right Click, another Republican-leaning tribe. (These people flocked in 1992 to the independent candidacy of H. Ross

Perot, a former high-tech entrepreneur. Their quieter cousins, the Civic Sentries, feel less economically alienated than the Right Clicks but worry more about safety and economic security.) Considering where we're sitting, in the liberal New Jersey enclave of Montclair, this man could conceivably be a moderate Inner Compass—someone who insists on fitness, both in body and balance sheet. He looks pretty buff, although with sweaters like that you never can tell . . . In the same swing group are those who care more about career satisfaction and material success: the Crossing Guards. Could it be? In the end, I can speculate until my coffee goes cold. If this were a fashion or technology study, my observations would inform me. I could document my conclusions. But to divine the political views of others—even those pecking away at laptops a mere eight feet away—is impossible. That is Gotbaum's refrain. Voters don't wear uniforms. Unless I go over to his table and lead him through the Spotlight questionnaire, there's no way of knowing which group he belongs to.

One blustery afternoon in New York, I ask Gotbaum if I can take the test. He's a little uneasy about it, worried that my knowledge of the process will skew the results. But he finally relents, and the next day he e-mails me a questionnaire. This isn't the 30-minute version, I'm disappointed to see, but a boiled-down sampler. In the course of their research, Gotbaum explains, the Spotlight team found 17 questions that predict, with 92 percent accuracy, how people would answer the others. So they cut to the chase. It takes me about five minutes to answer the multiple-choice questions about religion, schools, and my feelings about my neighborhood and country. When I get back the results, a day later, I learn that I'm a "Still Water." This is the less committed wing of the core Democratic group. It's the largest of the ten tribes and repre-

sents 19 percent of the voting population. Although 87 percent
of the Still Waters identify with Democrats, they're "independent-minded." I read on. They see "an affirmative role for government" but stand back from the "political vanguard." The
firebrand allies of the Still Waters are the Resourcefuls. They
have less esteem than Still Waters for entrepreneurs. Neither
group shows much interest in faith-based living.

I tell Gotbaum that he's not wildly off the mark. But as
I point out, I went to him and volunteered to be analyzed.
Most voters wouldn't dream of doing this. How can he determine which tribes the rest of the electorate fits into? How
can he find, say, 10,000 Barn Raisers in Tuscaloosa or Duluth?
Until he organizes the entire population into tribes, his political clients won't know where to start.

This is where consumer files came in handy, Gotbaum
says. To place all of us in our groupings (without having to
knock on millions of doors), the Spotlight team dredged Yankelovich's vast database of 175 million consumers. (That's 33
million more than voted in the 2004 elections.) Throughout
the summer of 2006, statisticians and data miners searched
for patterns within the files that would help slot practically
the entire voting population of the United States (along with
millions of nonvoters) into the ten political tribes.

To build our political profiles and sort us into tribes, they
harvested loads of tidbits with only the most tenuous relation
to politics. These are known as proxies, or stand-ins. They're
what statisticians rely on when they don't have the answers.
Here's an example. Imagine that you're catering a wedding
party and—horrors!—you've lost all the forms the guests filled
out with their dinner selections. You're too proud or scared to
take the simple route and ask people whether they prefer the
sausages or the vegetarian fricassee. So you study the people

and look for signs that might link each of them with one dish or the other. One man is telling loud jokes with what sounds like a strong Wisconsin accent. Wisconsinites, with their German heritage, eat loads of bratwurst, so you give him the sausage. You give the vegetarian dish to the skinny women, men with ponytails, a guy with a SAVE THE WHALES button on his lapel. Anyone who's overweight or drinking beer instead of wine gets the sausage. Judging people by these proxies is clumsy. Yet it's how most of us think. We analyze the patterns we know (or believe) to draw conclusions about others. The crudest of these form the foundations of prejudice and racism. Often they're wrong, or unfair. At this wedding party, quite a few of the guests will inevitably get the wrong dish. However, if you're smart in picking out these indicators, you're right more often than you're wrong. And if, like Josh Gotbaum, you have hundreds of data points on each person and a team of statisticians sifting through them, you can begin to place millions of us into tribes. (Usually we end up being grouped with others in our household, since most of our details, from the size of the mortgage to a *Field & Stream* subscription, are shared.)

So which nuggets of my data will expose me (and my wife) as Still Waters? Spotlight's statisticians can delve into all kinds of details. Cat owners, Smith tells me, are more likely to be Democrats. (We have two cats.) Republicans trend toward dogs. (We're down to zero, though I'm not happy about it.) Still Waters are more likely than their more ardent cousins, the Resourcefuls, to be college educated and married, with children living at home. (Yes, yes, yes.) We Still Waters show more interest in cooking than other groups do. A subscription to a gourmet magazine, Smith tells me, would help to identify me. (No such luck.) These details about pets, kids, cooking, and college education are proxies. They may have statistical relevance, but they're a far cry from evidence. On the other side of the scale, party registration is a clear political statement. (Like many independent-minded Still Waters, I don't belong to either party.) An even more compelling signal is the record of a financial donation to a political candidate. (My wife gives. If her donations are linked to me, I'm implicated, just as my father was by my mother's bumper sticker.) Reports of personal donations provide hard data. They leave behind the nebulous world of proxies. It's as if the wedding caterer saw the loud midwesterner wolfing down a plate of frankfurter hors d'oeuvres. Once the diner has established himself as a sausage eater, his accent and other statistical proxies are beside the point (though it's still possible that he ordered the vegetable dish).

As far as I can figure, unless the political data miners tie me to my wife's donations (she has a different last name but the same address), I blend into a large crowd of voters. That means they'll have to dig around for proxies. Which other ones tell the story? The traditional route is to focus on neighborhoods, ethnicity, income level, gender, and church attendance—all of the items that remained unchanged in my parents' case and failed to signal their political sea change. I return to Yankelovich's Smith. Which bits of data are most telling, I ask, these age-old variables or the far more exciting stuff about cats and cooking magazines? It doesn't have to be a choice, he tells me. He starts with the traditional groupings and then uses the newer data to highlight those of us who don't fit the mold. Of all the details in my file, the one Smith would seize on first is that I live in Democratic Essex County, New Jersey. Counties across the country are blue or red, he says, and that's a good place to start.

This is laughably crude. I tell him that while my county may vote Democratic, it's extremely diverse. It extends north from the rough Democratic wards of Newark, through liberal latte enclaves like Montclair, and into Republican suburbs. He responds that this is not the end of the analysis, but the beginning. For Democratic counties, Spotlight runs an algorithm looking for possible Republicans. These are people they're eager to bring back into the fold, or convert. In red counties, they plug in a different formula looking for signs of blue. In the early stages of this sorting process, they're hunting for exceptions to the rule. Those who look different are potential swing voters.

What makes them different? The data miners dig down into each county and take a look. Money is key. If our neighbors two houses down make three times as much as the average income on our block, or if they spent twice as much as I did for my house, they stand out. Why don't they live with their own kind? It might be a signal that they have different values. The Spotlight formula looks at their age and whether they have children living at home. "All these things have a bearing," Smith says. "Roughly 40 percent to 50 percent of the variance in people's values can be explained just by knowing their life stage and household characteristics." (And following my case, all of those details only confirm that my wife and I fit neatly into the patterns of a Montclair neighborhood teeming with Still Waters.) But as they dig deeper into the data, maybe they'll find something to set us apart. It's at this point that they start looking at more specifics—including the newer behavioral data.

I should point out that this protracted process, which might read like the work of a single analyst burrowing tirelessly through file cabinets, is carried out by a computer in the tiniest fraction of a second. It zips through our neighbors, our genders and ethnic groups; it looks at our magazine subscriptions, our credit ratings. It notes whether we live in a mobile home or a row house, whether we've ever taken a cruise. Altogether, it digests more than 100 different pieces of data for each voter. Bobbing up and down in that sea are clues that lead them—at least in theory—to profile each one of us as a political animal and to predict our behavior. They do scores of these analyses per second.

Let's look at how Spotlight might identify the techno-oriented Right Clicks, who account for 6 percent of the electorate. (Why the name Right Clicks? Folks who know their way around computers know all the extra tricks you can do by clicking the right side of the mouse. The rest of us click mainly on the left.) Right Clicks are the more committed half of the family-oriented segment, which they share with Civic Sentries. They lean Republican. But if a Democratic congressional candidate can come up with a list of, say, 10,000 of them, she can hit them with a direct mail appeal to their techno-libertarian values. It might stress, for example, that the government's broad surveillance of Internet communications is a Big Brotherly intrusion on our privacy—and that she would push for a more technologically sophisticated approach to finding terrorists.

The statistical method Spotlight uses to ferret out this tribe is known by data geeks as multiple discriminate analysis. Using the original test group, researchers create a model of a Right Click that they will then apply to the voting population at large. The model encapsulates a ranking of the details most likely to set each group apart. Assume, Smith tells me, that most of the Right Clicks in Spotlight's surveyed population are male, most have a broadband connection to the Internet, and most are white. Which of these three variables is most likely to distinguish them from the other tribes? Given the nature of the group, it's likely to be the broadband. Standing by itself, it's the roughest of predictors. Plenty of us have broadband. But the focus is on the statistical gap, or variance, in broadband subscriptions. How much more often do Right Clicks have them than the rest of us? Smith's team calculates that number and uses it to build the first piece of the model. The process continues. The team finds the second most important variable, and the third, and keeps feeding them to the computer. The researchers don't quit, Smith says, until they finally get down to a category—it might be the 50th, or 60th—that they deem statistically irrelevant. Maybe the fact that a particular voter is dogless doesn't make much of a difference. While the researchers have been introducing the ever less predictive variables, the machine has quietly been digesting all of the different probability rankings and turning them into a mathematical prototype of a Right Click. This is a preliminary model. Using it, the machine can theoretically sift through other consumer records and successfully pick out Right Clicks. The team tests it. If the model falls short and places too many Barn Raisers or Crossing Guards in the Right Click pool, they tweak and test again.

These models, when loosed, can sniff out voters everywhere. Imagine a pack of bloodhounds roaming America's cities, suburbs, and farmlands. Their heads aren't swimming with the scent of suspected murderers and rapists (those are sensory data). Instead they're packed with mathematical profiles. The neighborhoods they prowl exist in a database. Every time they pass the home of someone who appears to match one of these profiles, whether it's a conservative Bootstrapper (committed to individual initiative and "alloyed by a strong belief in a divine hand in human affairs") or the nostalgic Stand Pats (who long for a return to past values and feel that modern ambiguities "menace a lifestyle committed to patriotism, faith, family, community, and morality"), these hounds paw, whine, bark, or whatever it takes, in this matrix of numbers, to leave a record. (Good dog. Smart dog.) Yankelovich, in fact, has now run Spotlight's values models through every name in its jumbo database. So now, some 175 million of us are pegged as unwitting members of one or another of Spotlight's ten tribes.

As Gotbaum describes this method to me, I'm thinking far beyond these fanciful dogs. If political sleuths can build models of certain types of people, how far can others go? American jails and prisons house an ever-so-rich and varied population of criminals. As I write, other facilities, from the steamy barracks at Guantánamo Bay to holding tanks in the Middle East, house hundreds of suspected terrorists. What if researchers trawled through the personal data of convicted child molesters and then built a mathematical model of a pedophile? Would it be okay for schools or churches to screen job applicants by using this measure? If the tool had a proven correlation—say 50 percent, or 85 percent—would they be meeting their obligation to protect children if they ignored it? Would they be legally liable? What about those of us who are innocent but turn up as false positives in these analyses? Can we sue? As the Numerati advance, they're going to be measuring and profiling countless aspects of human behavior. This is going to raise torturous moral questions, ones that until now we never knew enough to ask.

Gotbaum tells me that his project was a success. His political sniffers, he says, managed to tag the three groups of swing voters with an accuracy rate of 75 percent. In many realms, getting one out of four wrong would be abject failure. But for a politician in these early days of microtargeting, reaching 7,500 voters with a targeted message is cause for celebration, even if the message spills over to another 2,500. That's a far higher hit ratio than broadcast TV can achieve. To reach my community in North Jersey with a political ad, for example, candidates often have to buy airtime on expensive New York stations. This means that their message spills to millions in New York and neighboring Connecticut who can never vote for them. They're also paying to reach loads of children, illegal immigrants, and the significant crowd of eligible voters who don't bother going to the polls. For campaigns accustomed to such staggering degrees of waste, reaching a targeted voter on three out of four tries sounds almost too good to be true.

Looking at it the other way, one quarter of us — 43.75 million American voters — are pegged to the wrong tribe. Gotbaum says that the errors put voters into a neighboring group. In other words, the system is almost right and doesn't mistake Bible-thumping conservatives for Communists. But still, what kind of science gets it wrong a quarter of the time? In two words: this one. Here, as on the other stomping grounds of the Numerati, the key is to forget about the truth — or at least put it to one side. While truth is vital and highly relevant in the world of machines (aviation engineers swear by it), the kind of statistical analysis we're discussing here, whether it's predicting our behavior as house hunters or wine shoppers, is by its nature approximate. It's based on probability. It involves all kinds of proxies in the place of real evidence. Truth is not a make-or-break test for the Numerati. They triumph if they come up with better, quicker, or cheaper answers than the status quo. Google, for example, doesn't provide definitive answers. It simply leads us to promising Web pages. In less than a second, it usually plunks us in the right neighborhood.

And because the earlier standard in Internet searching often left us lost and rudderless, Google rocketed to the top. Its approximations, crude as they were, turned a crew of algorithm-writing Numerati into a juggernaut.

The same goes in politics. Can the Numerati build models that connect candidates with voters at the right price? Are there areas where they can whip today's status quo, the precinct chiefs and TV advertising? Increasingly, both parties are concluding that the answer is yes. That's why political micro-targeting—the domain of the Numerati—is the latest rage.

IN THE EARLY DAYS of 2001, President Bush's chief political strategist, Karl Rove, was still wondering what had gone wrong. Going into the last 72 hours of the previous November's elections, the Bush team had been leading in all the polls—and yet Al Gore had won the popular vote. And he had come within a few dangling chads and one Supreme Court vote of winning the whole shebang. How could the Bush team guarantee that this wouldn't happen again?

Rove gathered strategists over the following months into the so-called 72-Hour Task Force. Later the task force condensed their conclusions into a PowerPoint presentation that circulated within the Republican Party. It called for all kinds of improvements, from getting out a coherent message to re-cruiting election-day volunteers. But one of the most promi-nent was microtargeting. "It's the oldest adage in advertising," say the notes accompanying the slides. "It is always easiest to sell people what they want to buy." To that end, Rove's task force urged party activists to "take every list you can get your hands on, and add the information to your voter file. This can be everything from lists of realtors, lists of chamber of com-

merce members, church directories, professional associations
. . . We must adopt the idea that no list is too small."

In *Applebee's America,* Bush's campaign strategist Dowd
and his coauthors detail the Republican approach. The goal
was to map the political DNA of voters in Michigan, an important swing state. Like Spotlight, the Bush team combined
surveying with large consumer-behavior databases. But the
approach was different. The questions hewed much closer to
political issues. Instead of searching for core values, the team
measured responses to political issues that had already arisen
in public debate. It was just a matter of figuring out which
ones moved them. Were voters upset by the prospect of gay
marriage? Did they fear terror attacks? Were they outraged
by smog or child porn on the Internet? When the team got
through the surveying, they combined demographics and survey data to create 31 finely tuned segments, such as Terrorism Moderates and Middle-Aged Female Weak Republicans.
Then they used every bit of data they could get their hands
on, from magazine subscriptions to voting records, to peg the
state's voters. "If John Doe earned $150,000, drove a Porsche,
subscribed to a golf magazine, paid National Rifle Associa-
tion dues, and told a Bush pollster he was a pro–tax cut conservative who backed President Bush's war on terrorism, the
Bush team figured that anybody with similar lifestyle tastes
would hold similar political views," according to Dowd and
his team.

Let's imagine a voter who's harder to pigeonhole than that
gun-toting suburbanite. Maybe he lives on the very same cul-
de-sac and makes piles of money, but he voted twice in the past
decade, according to the records, in a Democratic primary.
And he drives a ten-year-old Subaru, a liberal car if there ever
was one. Hmmm. Political analysts increasingly look at voters

and give them numerical grades, much like Fair Isaac's credit-risk scores. In a 2005 governor's race in Virginia, for example, every single voter in the state received a "likelihood score," from zero to 100, that he or she would vote for the Democratic candidate, Tim Kaine. The inscrutable Subaru driver I just described might rate a 50. This scoring system made targeting easy. The Kaine campaign wrote off the voters with low scores. And they hardly bothered with those registering scores of 90 or above (except as potential donors). That would be preaching to the choir and a waste of resources. Instead, they focused on promising swing voters, those with scores from 55 to 75. "If you were a 60, you were getting communicated with. We were all over you," recalls Kaine's victorious campaign manager, Mike Henry.

But what message were they delivering to that pool of swing voters in Virginia? Josh Gotbaum would argue that those voters represented a big stew of Barn Raisers, Civic Sentries, and Hearth Keepers, garnished perhaps with a few Right Clicks and Inner Compasses. His approach would call for each group to receive a different stream of mail and telephone calls. Even the same issue—raising the minimum wage, for instance—would be framed differently. Community-minded Inner Compasses might hear that their neighbors needed a decent wage to live a healthy life, while the more conservative Civic Sentries would learn that a higher minimum wage would give hardworking families what they needed to make it on their own. In the Virginia race, Kaine's team had to pick issues that would appeal to all of these swing groups. Following a large voter survey, they focused on better schools and wider roads. But that was 2005. Henry says that in coming elections, the targeting will be far more sophisticated. He and others—especially those working for Democratic presiden-

tial nominee Barack Obama—are gearing up for an unprecedented explosion of political warfare fueled by data.

They'll be wrestling with layer upon layer of statistical
complexity far beyond the tribes I've mentioned. Think for
a second about one of those Virginia voters. How much is
a 90 worth if he votes only once a decade? How about a 55
who braves blizzards and floods to make it to the polls? Two
variables, level of support and likelihood of voting, are both
crucial. And now the political Numerati are reaching into the
tool kits of economists to calculate a projected rate of return
for every advertising and promotional dollar spent on each
one of us. In other words, how much will it cost to turn you
into a vote for their side?

"I was working with a theoretical economist I went to
graduate school with," Mark Steitz is telling me. Steitz, a longtime Democratic consultant, operates out of a townhouse just
up Connecticut Avenue from the cafés and bookshops of Du-
pont Circle. "We started thinking abstractly about the best
way to formulate this problem," he says. "And we came up
with this triangle." He clicks his computer, and a red-and-blue
image appears on the screen. This so-called simplex triangle
represents the universe of voters in an election. The position
of each voter on the triangle is determined by two calculations: the probability that the voter favors Republicans or
Democrats, and the probability that he or she will vote. Steitz
draws a vertical line up the triangle, a so-called isoquant. Each
voter along this line is of equal value, he says. A voter who
leans to the Democrats 75 percent of the time and votes in
every election is on the same isoquant—and has the same
value—as a voter who's 100 percent Democrat and votes three
out of four elections. Those two voters, Steitz says, "are indistinguishable to me." His triangle, at this stage, is largely theo-

retical. Yet as politicos learn more about voters, they'll be able to plug us ino more of these types of mathematical formulas.

As this happens, the calculations grow only more complicated—a trend that plays to the strengths of the political Numerati. Some votes, it turns out, are worth far more than others. Each side in an election needs only 50 percent of the votes, plus one. That one vote at the end could be worth millions of dollars. Just think of the handful of contested Florida votes in the 2000 election between George W. Bush and Al Gore. And yet a vote that lifts a candidate to 60 percent, or to 40 percent, has only symbolic value. And that last wavering voter, according to Steitz's triangle, will be the most expensive to acquire. As the Numerati develop tools to model voters and measure the effectiveness of campaign spending—its "yield," in economic terms—political parties will be able to look at each election as a marketplace. As the polls swing, the relative value of each voter rises and falls. Some of us are cheap, virtual throw-aways. Some will be prohibitively expensive, not worth the investment. But those of us who shape up as the difference makers will be the targets of increasingly customized come-ons. Analysts will know which of us are wrestling with college tuition and which of us fear that our jobs will leave for India. Some might even voice a concern about an outbreak of rabies that threatens our cats. If the politicians get it right—which is no sure thing—the campaign messages will address our concerns and reflect our values. It'll be as though they finally understand us. Who knows? Maybe they'll even learn not to call us at dinnertime. We'll feel, if only for a few short weeks of a frantic campaign, appreciated.

CHAPTER 4

Blogger

"NOW LET ME tell you of my long tale which I have taken time to tell." These words draw me in. I sip my coffee and scroll down the page.

It's a long blog entry written by a woman who calls herself "Tears of Lust." She writes about driving all night to Columbus, Ohio, with her boyfriend, Kenny, and her sidekick, Lizzy. The three of them are headed to an anime convention. As they drive through the night, Kenny starts to feel sick. By the time they reach Columbus, he's "passed out on the [hotel] bed."

Kenny ends up in Columbus's Grant Medical Center and faces an emergency appendectomy. "Kenny made his phone calls to relatives before he went under," Tears of Lust writes. "And me and Lizzy went to the waiting area. We watched *Kill Bill* and read a magazine to each other before returning to the café which was closed but they had the best vending machines like the ones that rotated and the ones that were refrigerated. It was awesome."

The story goes on. It's a medical saga woven with the wanderings and cravings of a consumer. "At one point," she writes,

"[Kenny] thought he heard his dead aunt talking to him so he assumed he was going to die during surgery." But Kenny survives, and he recuperates as the writer and Lizzy happily watch *Memoirs of a Geisha* and *The Da Vinci Code* on DVD. Later, we learn some details about his stomach problems, his nearly collapsed lungs, and an abscessed wound. He's so tender that when he and his girlfriend make love a few days later, Tears of Lust has to climb gingerly on top, "like a naughty nurse."

That's just a quick dip into the blogosphere, an enormous and expanding pool brimming with some of our most personal data. Up to now, we've seen how employers can track our procrastination and our e-mails, and how they'll be able, increasingly, to optimize us as workers. We've seen how advertisers attempt to turn our mouse clicks and movements into mathematical models that anticipate our every urge. In what we've seen so far, it's others who have their way with our growing mountain of data. They grab it, they analyze it, they use it. Whether we're shopping or taking out a loan, we're laboring for the Numerati in much the way a drosophila fly works for a white-coated lab technician. Sometimes we get discounts and prizes. Sometimes we can say no. But once we agree to an offer, we're specimens. And yet, in the world of blogs and YouTube and social networking sites like MySpace, millions of people broadcast their lives voluntarily. They pile up details by the shovel load. Privacy often looks like an afterthought, if it's considered at all. People like Tears of Lust aren't pawns. They're running the blogosphere. But that doesn't mean they can't be used.

On a frosty winter morning in New Jersey, I take the laptop to a coffee shop and call up Technorati, a blog search engine. There I look for a post brimming with the kind of private details most of my friends and acquaintances would rather keep to themselves. To limit the field to informal writers who hold nothing back, I type a misspelled "diahhrea" in the search box. The first post that pops up is by Tears of Lust.

For market researchers, blog posts like this one open a window onto a consumer's life. Blogs and social networks offer up-to-the-minute intelligence—something marketers have long dreamed of. For decades, soap makers, brewers, and movie studios have attempted to simulate the marketplace, at great expense, by bringing together focus groups. These small groups of people, usually numbering from eight to a dozen, agree to try out the latest jelly beans or toothpaste, watch competing ads, or view a Hollywood release. Marketers watch to see if group members squirm or yawn during horror movies, if they nod or recoil while watching a political attack commercial. They have to make the most of each gathering because putting together focus groups is expensive and budgets are tight.

Now that people like Tears of Lust are publishing their feelings about a host of products, it's as if a universe of focus groups is forming online. Tens of millions of people participate. Many write copiously. And from a marketer's view, many are gloriously indiscreet about practically everything. True, some of them, like Tears of Lust, shield their identity, or at least change their name. Marketers don't care. What they relish is the unfiltered peek at the moving gears and conveyor belts of peer pressure, bias, and desire.

Bloggers tend to be younger than the average consumer and a bit more tech savvy. Statistically speaking, they don't reflect society at large. Still, it's a big and surprisingly diverse pool of people, numbering more than 20 million. Grandmothers blog. CEOs blog. Marketers can dive into these online journals to find opinions about nearly everything and to track trends. The only trouble is that no one employs big enough teams of readers to keep up with the blogs. No one could. It's too much text for human eyes. And the themes of the blogs, much like our lives, wander all over the place. They're hell to organize. The only way to harvest and file the customer insights streaming from blogs is to turn over the work to computers.

IT's A WHITE winter in Colorado. Every weekend, it seems, another blizzard blows in. Little surprise, then, that I get a fabulous rental deal on a convertible. The wind whips the cloth top as I drive it west through Denver and up toward the snowy mountains, to the university town of Boulder and the home of Umbria Communications. It's a company that mines the millions of words pouring into blogs every hour. Its purpose is to learn what you and I and everyone else in the consuming world are thinking and, especially, what we're hankering for.

Howard Kaushansky, Umbria's president and founder, got my attention early on when describing Umbria's business. "We turn the world of blogs into math," he said one day, while visiting New York. "And then we turn you into math." A colleague and I had just launched our own blog. The idea of turning it into math sounded like a lot of work. And turning me into math? I supposed such a thing was possible, but I had only a vague idea why Umbria's team would bother and no clue as to what kind of equation I would become. Truth be told, as I drive into Boulder, I'm still struggling with the concept. I've made my way through entire chapters on hidden Markov models and Bayesian analysis. I've even wrestled briefly with so-called support vector machines. But I'm still not sure what you and I look like in numerical form. That's one of the things I've come to Boulder to learn.

Kaushansky has smooth rounded features on a wiry frame, neatly coifed graying hair, and the restless mien of a marketer. In fact, he's a lawyer by training, and he's been running businesses in analytics and data mining for the past 15 years. He builds and sells them. The last one was Athene Software, a predictive analytics company he sold in 2001. With Umbria, he's focusing the analytics on blogs.

A bicycle wheel leans against the wall in Kaushansky's office. I ask if he rides. He does, he says, brightening. Extensively. I do too, I tell him. The area around Boulder is a "dream" for cyclists, he says. (I hold back from my spiel about the unexpected biking jewels in New Jersey. People don't care.) I ask him where he lives. He points out the window toward an anvil-shaped mountain, Flatiron. He lives on the other side of it. He sees bear out there, herds of elk some mornings, coyotes. Leave your domestic pets outside after dark, he says, and they'll get devoured. He spends weeks on end away from this mountain spread, flying around the country. He's trying to convince nearly every type of company to tune in—through Umbria—to what their customers are saying on the blogs.

Kaushansky founded Umbria in 2004. Since then, the company has built a system that automatically reads millions of blog posts every day. The first step, he tells me, is to learn something about the author of each blog. Is the person a male or a female? A teenager? A twenty-something? A boomer? The computer looks for clues such as sentence structure, word choice, quirks in punctuation. How many middle-aged men do you know, for example, who end a sentence like this: !!!!!!!!!!!!!!!? Sometimes the computer reads through a post, shrugs its digital shoulders, and gives up. It doesn't see the tell-tale signs. That post goes unclassified. Despite such setbacks, says Kaushansky, Umbria's computer is able to build up large piles of posts for each gender and generation. They sort the authors into those categories.

The next step is to figure out what each group of authors is writing about. In a decade or two, automatic readers like Umbria's will likely dive deep into the content of written documents, perhaps analyzing an author's mood, income, and educational level. Maybe the computer will come to conclusions about the individual's circle of friends or be able to predict his or her behavior. For now, though, with only a tiny sliver of a second to devote to each blog post, Umbria is delivering far simpler fare. It's looking to see if the writers have opinions about services or products—a new cell phone, for example, or the call center for a large bank. The only conclusion it reaches is whether the blogger has a favorable or unfavorable opinion. Thumbs up, thumbs down.

It sounds crude. What makes the blog world especially valuable to marketers, though, is not its precision but its unfiltered immediacy. Opinions change day by day, sometimes hour by hour. Let's say that one of Umbria's customers launches a new deodorant on Tuesday and promotes it with $4 million of TV advertising over the following week. How can marketers find out if the advertising has reached the target audience? Most of us don't rush out to buy deodorant, no matter how compelling the ad. We might have another two or three months left on the stick in the medicine cabinet. So sales figures won't provide quick feedback. Traditional Web pages, the kind that search engines like Google comb through, are static, a bit like a library. They're sorted by relevance, not timeliness. Chances are, the most "relevant" Web page is the company's own press release. That doesn't help one bit. To learn what we're thinking, the deodorant company must reach beyond the more formal Web to what bloggers and social networks are saying about the product.

This may sound like an outlandish example. People blogging about deodorant? But now that every single person online can become a global publisher in the five minutes it takes to set up a free blog, the sorts of details that people publish might surprise you. I search on Technorati for "deodorant," and within minutes I find a post from Jeff, a 46-year-old "ex-touring musician turned husband/dad," who lives in St. Cloud, Minnesota. He takes us on a tour of his bathroom cabinet, presenting his views on everything from floss ("It has to be the really thin stuff since my back teeth are very close together") to mouthwash ("I've never been a mouthwash kind of guy. And that disgusting commercial where that guy swooshes hot, spitty mouthwash around . . . for ten minutes doesn't help much either"). He informs us that he stopped buying cologne after marrying, since he no longer needed "to wear bait." And yes, he weighs in on deodorant: "If any of you have teenage boys, I hope for your sake they don't find out about Axe, or any of the other popular body sprays. My wife and I have had to intervene to let them know that they don't need to spray on a whole can at once when a half a can will get the job done just fine. Ugh."

By rounding up this kind of consumer insight, Umbria can provide the advertiser with a report showing how much buzz their ads generated the first day, or the first week, of the campaign. It can determine whether the response was favorable and how it matched up with the competition. (In this example, the demographic details are crucial. If the company is marketing the deodorant to Jeff's teenage kids, the "Ugh" from their father might not even be a negative. Jeff makes it easy for Umbria's computer by putting his age and gender on the blog. (We even learn that he's a Leo.) This type of research turns traditional surveying on its head. Unprompted by marketers, bloggers like Jeff volunteer the answers to millions of potential questions. "In a sense, we're very similar to the game show *Jeopardy!,*" Kaushansky says. "People have already said that they like a certain car or dislike a movie. It's our job to formulate the questions."

Kaushansky's team is also starting to divide bloggers into different groups, or tribes. Kaushansky envisions nearly endless tribal affiliations. Doritos munchers, bikers for Obama, MINI Cooper enthusiasts. Once the company has sorted bloggers into tribes, it can start digging for correlations between tribes and products. It was through analysis of blogs, for example, that Kaushansky learned that the Gatorade tribe includes not only athletes and fitness nuts but also heavy drinkers on college campuses. Many use it as a mixer in hopes that electrolytes in the drink will soften hangovers. If this was news to Gatorade executives (and to their credit, it wasn't), the company could consider extending its promotional partnerships beyond the likes of Nike and Cannondale, perhaps reaching out to Bacardi or Absolut vodka.

Sometimes this tribal knowledge helps marketers draw distinctions between consumers. One cell phone company, Kaushansky tells me, started charging extra for Bluetooth data connections — radio signals that replace wires. The people who wear the blinking phone clip on their ear use Bluetooth to relay their conversation to their handset. News of the extra charge on the phone bill sent bloggers into a fury. But Umbria, Kaushansky says, studied the blogs and discovered that almost all of the anger came from one tribe: the "power users." Those are the folks who spend lots of time and money hunched over their handsets, sending e-mails and photos and fooling around with spreadsheets. The other tribes—the fashionistas, the music lovers, the cheapskates—shrugged off the Bluetooth charge. Many of them probably didn't know what it was. With this intelligence, the phone company, conceivably, could raise the price a few bucks on the handsets favored by power users, and then offer them "free" Bluetooth. Meanwhile, they could continue charging everyone else.

This is a new stage in market intelligence. While still primitive, it's easy to see where it's headed. The Numerati are training computers to digest our words automatically. They are coming to conclusions about who we are and what we think. And as these computer systems gain in speed and ability, they'll feast on more of our communications, extending far beyond blogs. Automatic readers like Umbria's can stretch into social networks like MySpace and Twitter, the meeting places of entire generations. They can surf the comments on interactive video games and dig deeper into our e-mails, picking out our hobbies and passions and selling that information to advertisers. Given technology like Umbria's, scores of companies with access to our words will be positioned to track, minute by minute, the shifting patterns of human thought. Umbria and its competitors, from Nielsen BuzzMetrics to Google, are betting that marketers, government officials, and politicians will pay richly for the insights they come up with. And this analysis of our words may be speeding ahead even faster in the shadows. Following the terrorist attack of 2001, intelligence officials in the United States gained access to enormous flows of Internet and telephone traffic. The National Security Agency, which has the largest staff of mathematicians in the country, is mining the traffic hour by hour.

Until recently, our words, whether spoken, typed, sung,

or scrawled, performed their magic beyond the range of mathematicians. It wasn't just that language, with its countless shades of tone and shifting nuances, resisted the rigid hierarchies of geometers and computer scientists. (That's still an issue, as we'll see.) No, the problem was more fundamental. Our words didn't hang around for analysis. The sentences we spoke traveled through the air or along copper wires before alighting, every so briefly, in forgetful minds. They faded faster than cut flowers. Our written words moldered on pages, a select few of them stuffed away in envelopes and notebooks. Most of them weren't in the public domain, much less on the hard drives of powerful computers.

That has changed. For starters, our queries to search engines provide a detailed timeline of what online humanity is interested in—what we're looking for, what we would like to buy. But those queries, most of them just three or four words long, are bare bones. They point in a direction but divulge only odds and ends about the people who write them. Think about what you searched for online over the past week. Those queries might trace your pursuit of a high-definition TV or your research for a geology project on the Pleistocene era. But they could easily miss important events in a person's life—the death of a parent, perhaps, or a battle against addiction. Outfits like Umbria are working to glean new marketing insights from these online rambles. Imagine that they might want to create a bucket of several thousand female bloggers trying to quit smoking. It wouldn't be hard. Now, do they appear more interested than average in chocolates or in white wine? In these early days, Umbria is focusing on simpler stuff. But a wide-ranging sampler of human life is circulating on blogs, ready to be harvested. It's as if humanity itself were squeezed right into Umbria's offices, sheltered from the winter winds outside, typing on command. Once the words show up, they're available for eternity, to be matched, compared, crunched, parsed, and repackaged as marketing intelligence.

Maybe you don't write a blog and you stay off social networks. If so, you could be lulled into thinking that companies like Umbria learn about other people, not you. But that's not exactly the case. Umbria and other analytics companies are going to school on blogs. Once this new generation of automatic reading and learning machines scopes out the blog world, they can broaden their focus to everything else we write. In fact, it's already happening. Spam-fighting companies such as Postini—a division of Google since mid-2007—sift through millions of e-mails issuing from Fortune 500 companies. They check them for signs that employees are leaking company secrets or carrying out insider deals. Other companies sweep through the hard drives of corporate flotillas of personal computers, scanning words to make sure that employees aren't using the equipment for their own sordid or selfish ends.

Companies and governments alike are poring over our written words. Most of the snooping is focused on crime prevention. But as the tools improve, the market will change. Instead of just looking for what we're doing wrong, companies and governments and pollsters will be eager to learn what we're buying, where we're going, who we might vote for. They're curious. As Umbria and others gorge on data in the blogosphere, they are sharpening the interpretive tools.

I READ Tears of Lust's blog, and it's easy to figure out quite a bit about the writer. She's a young woman. She lives in a city on the East Coast of the United States. If I had to guess, I'd

say New York. But I can't bet on it. I could come to lots of
other conclusions about her passions, her love interests, and
even what she likes to eat.

This is all clear to me. But she's writing in my language.
Practically every word makes sense. The bad news, from a
data-mining perspective, is that it takes me a scandalous five
minutes to read through her text. In that time, Umbria's computers work through 35,300 blog posts. This magic takes place
within two domains of artificial intelligence: natural language
processing and machine learning. The idea is simple enough.
The machines churn through the words, using their statistical
genius and formidable memory to make sense of them. To say
that they "understand" the words is a stretch. It's like saying
that a blind bat, which navigates by processing the geometry
of sound waves, "sees" the open window it flies through. But
no matter. If computers can draw correct conclusions from
the words they plow through, they pass the language test. And
if they can refine those conclusions, adding context and allow-
ing for caveats, then they're getting smart — or, as some would
have it, "smart."

For decades, computer scientists have battled over how to
teach computers language and thought. Some have pushed
for a logical approach. They follow a tradition pioneered by
Aristotle, which divides the entire world of knowledge into
vast domains, each with its own facts, rules, and relationships.
One of the most ambitious of these projects, Cycorp of Austin, Texas, is trying to build an artificial intelligence that not
only knows much of the world's information but also can
make sense of it. When you ask Cycorp's computer for dem-
ocratically elected leaders in the Northern Hemisphere, the
geography arm of the system starts whirring away, country
by country: Britain is in Europe. Europe is in the Northern

Hemisphere. The Northern Hemisphere is north of the equator. It knows each of these facts and skips from one to the next. Then, according to Cycorp's Web page, it uses logic: if region A is part of region B, and region B is north of region C, then region A is north of region C. Therefore, it concludes, Britain is north of the equator. At that point, the geography arm has to ask its political cousin if Britain is a democracy. And the analysis continues. The trouble with this logical approach is speed and flexibility. Facts change, challenging the immense system to rejigger its bits of information and the relationships among them. In 1984, when Cycorp began assembling its knowledge universe, the Soviet Union dominated the Asian continent, and the "mouse" was barely pushing its nose from the domain of rodentia, a subgroup of mammals, into the burgeoning arena of computer accessories.

The rival approach rejects this plodding logic and prefers to see the computer purely as a counting whiz. Statistics are king. Probability defines truth. Speed and counting trump knowledge, and language exists largely as a matrix of numerical relationships. This is Umbria's tack—and it is this statistical approach that most Numerati hew to as they study us in nearly every field. What the computers in Boulder learn is a crazy quilt of statistics and geometry. They may come up with startling insights. But they reach them through a labyrinth of calculations. These learning machines swim in numbers.

The learning process starts with humans, a team of 6 readers at Umbria headquarters and 25 colleagues in Bangalore, India. These are the annotators. They go through thousands of blogs manually, looking for evidence of each blogger's age and gender. Is the person male or female? If so, how old? Sometimes they can't tell. But when they can, they mark the blogs and put them into a digital folder. Their work is to build a "gold standard," a selection of accurately labeled blogs that can be used to teach the machine. In this process, Kaushansky says, Umbria researchers put perhaps 100,000 blog posts in the golden folder. They take 90,000 of them and introduce them to the computer. They leave the other 10,000 to one side.

How do those annotators come to their conclusions about the blogs? In many cases, they rely on knowledge and context that would be hard to teach a machine. I pretend I'm an annotator and reread Tears of Lust's post. From the very first paragraph, I'm convinced that it's a woman writing. But what tells me this? It's a tone I pick up, a voice. These are hard things to enumerate, never mind impart to a machine. Telling details? There are a few, but they're not definitive. A man could have a boyfriend named Kenny. A man could get dressed up with Lizzy and search for thrift stores. I suppose a man could even write "I did a good job getting around Columbus and know the city pretty well. Go me!" It isn't until later in the post that I read, "So as brother, sister, and fiancée we went up to surgery where we met with plenty of doctors." It's a little unclear who's who. But I review the section until I'm just about sure that Kenny's the brother, Lizzy's the sister, and Tears of Lust is the fiancée. If I were an annotator, at this point I would confidently mark her blog post F, for female.

A computer, Ted Kremer tells me, would have to look for other signs. It wouldn't recognize the author as the fiancée, and it might not know that a fiancée is a woman. Kremer is the chief technical officer at Umbria. He works in a huge sunlit office lined with whiteboards. He's blond, with a square face, a pointed goatee, and near limitless patience—at least when it comes to teaching basic data mining. Once the annotators have created their golden files, he tells me as he scrawls on a whiteboard, the scientists comb through the documents, looking for hundreds of variables the computer can home in on. They look for telling words and combinations of verbs and objects. ("Go me!" might be one.) They look at punctuation, at certain word groups, at the placement of adjectives and adverbs. They adjust for certain groups of words that, taken individually, could be misunderstood. The computer has to know, for example, that Denver Broncos are a football team and not, Kremer says, laughing, "A city of horses." His team also looks at strange spellings. Some bloggers spell *great* as *gr8*. Others sprinkle their posts with online gestures known as emoticons, such as the smiley face, :). The scientists instruct the computer to keep an eye on the fonts used and the color of the blogs' lettering and background. (Tears of Lust, I note, uses a blue background covered with pictures of faces, making it look like a Bollywood movie poster.) Add it all together, and the computer might have more than 1,000 different features to find and count.

Can the computer distinguish between the sexes? The test is ready. The computer plows through the 90,000 blogs at lightning speed. It counts every one of the variables, and it arranges them by gender. Kremer's team studies the results. Are there certain quirks or features that are far more common in one gender's posts than the other's? That's what Kremer's team is hungry for. Correlations. If they find them, they assemble them into a model. This is the set of instructions that tells the computer how to tell a male blogger from a female one. It starts off with the easy stuff. Some bloggers, for example, identify themselves as M or F at the top part of the blog known as the header. That's close to a sure bet. Certain phrases are strong indicators: "my dress," for example, or "my beard." But most of the model is a statistical soup of more subtle signals, of varying verb combinations, punctuation, and fonts. Each one is tied to a probability. The art in this science involves how much weight to give each component.

When the scientists have their model ready, they try it out on the 10 percent of the gold standard documents they had put aside. They quickly see how many of the documents the model sorts correctly and, more important, which ones it misses. They pore over these, looking for signs of faulty analysis. Why did it screw up on one man's post? Did it give too much weight to the girlish exclamation points? Did it overlook the phrase "Guys like me"? How silly. But then again, that phrase may not have popped up in the first test batch, so it wouldn't be in the model.

Like a chef whose soufflé comes out too salty or flat, the scientists adjust the variables. They tweak their model. They take weight away from some components and add it to others. A few new hints may have popped up in the latest sampling. Those can be factored in. This process goes on and on, sometimes through ten iterations or more. In this sense, the computer is a maddenly slow learner. It takes an entire laboratory of scientists, often skipping the slopes on weekends and ordering pizza late at night, to teach the machine what we humans know at a glance. And when the computer finally passes its gender test (Umbria won't disclose the accuracy rate), the machine is still a long way from graduation. It moves on to the next stage, learning to peg each writer to a generation. Here, some of the markers are surprisingly simple. Older writers, for example, use a greater variety of words than younger ones do. Counting such words is hardly foolproof, but it gives the Umbria system a running start in generational groupings. It isn't until the computer has worked through gender and age that it faces its most difficult assignment: figuring out whether bloggers give a thumbs-up or thumbs-down to whichever food, soda pop, music, or political candidate they're analyzing.

This is a tortured way of learning about us. Instead of knocking on our door, data miners break our documents into thousands of components and then sift through them obsessively, attempting to put together a mosaic of our thoughts and appetites. It's sneaky. It reminds me of parents who, instead of asking their teenager point-blank where he drives at night, tiptoe into the garage, take readings from the odometer, and make their own projections on a map. This approach is often less precise and a whole lot more work. But data mining plays to the counting and calculating genius of the computer, and above all to its speed. And it takes a wide detour around the computer's weak points—specifically, its limited ability to think and to understand.

I ask Kremer about this. It seems to me, I say, that a computer working from a set of instructions, no matter how exhaustive, must make loads of mistakes. After all, we humans —each of us carrying a prodigious brain wired for communication—misread each other's words and gestures daily. "What?" we say. "Huh? Are you kidding? Oh, I'm sorry, I thought . . . No, what I meant . . ." If you listen, we're constantly amending what we say. Getting our meaning clear in someone else's head and understanding what they're trying to say is a struggle to which we devote an enormous amount of our intelligence. Entire industries, from psychology and law to literature, focus on sorting us out. They never run out of work. So computers, I ask Kremer, poor dumb computers that can count a million trees without realizing that they're dealing with a forest, they must sometimes get utterly confused. Right?

Sometimes, he says. He leads me over to his computer, and we look at a blog post that's been analyzed by Umbria's system. It's an article about Apple's iPod Shuffle. Phrases in red are deemed negative. Green phrases are positive, and blue unknown. I look for a section in red. One of them reads: "Jobs not only moved the exalted Shuffle to the top of the menu . . ."

I study the phrase and see nothing negative about it. Could *exalted* be sarcastic? It doesn't look like it. I call Kremer's attention to the section. He reads it and shrugs. "False positive," he says. "It happens." He points to the "not" in the sentence. That simple negative may have led the computer to see the entire phrase as a condemnation.

Sarcasm, Kremer says, stumps the machine on a regular basis. It's probably okay to take a San Diego blogger at her word when she exclaims, "I LOVE THIS WEATHER!" Yet how can the system know that when a blogger 1,000 miles north, in soggy Portland, writes the very same sentence, *love* may very well mean "hate"? These are the challenges of machine learning, and they fuel graduate research in top programs around the world. Tackling sarcasm may involve teaching the machine to keep an eye out for messages in capital letters, exclamation points, and the tendency of teens to indulge in it more often than their grandparents do. Perhaps in the distant future, contextually savvy machines will be armed with a long list of meteorological gloom zones. Maybe they'll "understand" that *salt* in such latitudes often refers to highway issues, and not cuisine, and that raves about the weather, at least during certain seasons, are bound to be facetious. But for a company selling services today, such exercises are academic.

"I've got bigger fish to fry," says Nicolas Nicolov, Umbria's chief scientist. A Romanian-born computer scientist, Nicolov got his doctorate in Edinburgh before moving to America, first to IBM's Watson lab and then to Umbria. He has an angular face and dark deep-set eyes, and he sports thick black bangs—a bit like Jim Carrey in his early movies. He works in a small, dark office down the hall from Kremer's sunny, expansive digs. It feels like I've stepped into a cave.

Nicolov gives me an example of the kind of confusion he has to sort out. Umbria does lots of work for consumer electronics companies, he says. They want to know what kind of buzz the latest gizmos are generating. But in this area, even words like *big* and *little* vary with the context. "If a laptop is big, it's negative," he says. "But if a hard drive is big, it's good."

Nicolov and his team can teach these lessons to the computer. It helps enormously to train it for a specific industry, or what computer scientists call a "domain." Within that area, it learns not only words but also groups of words. Bigrams are pairs, trigrams are triplets. Anything bigger than that is an Ngram. So a sophisticated machine trained for laptops might draw a green line under a trigram like "big hard drive." That's positive. But it might not be quite so confident when confronted with the Ngram "big honking hard drive." That one it might miss.

Do these errors skew Umbria's results? Kaushansky maintains that he's chosen precisely the right market for inexact results. "We're providing qualitative research, not quantitative," he says. "It's directional. It gives early indications of where things are going, what new issues are popping up for a company." To make his case, Kaushansky shows me Umbria's tracking of President Bush during his 2004 reelection campaign. He has the favorable and unfavorable blog references to the president on a chart alongside a series of Gallup polls. The Umbria blog numbers appear to anticipate by two to four weeks the ups and downs in the polls. Kaushansky is saying, in effect, that even if his computer misinterprets the words on our individual blogs, it reads our trends. It tracks our tribes.

But who exactly populates those tribes? It's a burning question among blog analysts. The tribes, after all, are defined not by neighborhood, race, tax bracket, or the answers checked off on a survey. Instead, machines analyze our words and drop us into tribes with people we might be surprised to encounter. These tribes are a little like the buckets we land in at the supermarket—but with an extra layer of complexity. At the grocery store, consumption patterns are all that counts. But Kaushansky's tribes, like Josh Gotbaum's political groupings, have to embody an entire set of related values.

Kaushansky gives me an example. Four years ago, a 43-year-old friend of his rediscovered his adolescent passion for skateboarding. He's a true fanatic, Kaushansky says, and he adores not only the skateboard but the whole culture that surrounds it. He talks like a teenager, in Kaushansky's view. He listens to a different generation of music. And here's the important part: Kaushansky insists that he writes like a skateboard-obsessed teen on his blog. Perhaps in a decade or two, systems like Umbria's will be able to distinguish between true teens and middle-aged poseurs. But not today. In their statistics, this 43-year-old is likely to show up as a teen. He aches to be a member of that tribe, Kaushansky says. And for Umbria's purposes, what difference does it make?

A few weeks later I'm in the San Francisco offices of Technorati, and I relate the story of the skateboarder to David Sifry, the search engine's founder. Sifry, a transplanted New Yorker with not an ounce of West Coast cool in his expansive body, explodes: "WRONG! WRONG!" A man can write like a woman, he says, but does he buy like a woman? Sifry goes on at length about the dangers of predicting people's behavior based on statistical correlations. "Let's say that according to my analytics, you said that *Mission Impossible III* was no good and that you can't wait to see *Prairie Home Companion*," he says. "I can't assume from that that you're an NPR listener. That's where you get into trouble." That's mistaking correlation for causation, he says. It's common among data miners—and most other humans. How many times have you heard people say, "They always do that . . ."?

For Kaushansky, putting his skateboarding friend and a few others in the wrong tribes may not turn out to be too serious. That's why advertising and marketing are such wonderful testing grounds for the Numerati. If they screw up, the only harm is that we see the wrong ad or receive irrelevant coupons. However, as the Numerati file into other industries, such as medicine and policing, they won't have the luxury of tossing loads of us, willy-nilly, into the same piles. Instead of concentrating on what we have in common, they'll have to search out the data that sets us apart. It's a much tougher job.

EARLY IN 2005, blogging was developing into a craze. Political bloggers had swung their weight in the 2004 presidential campaign, and now some 40,000 new bloggers were popping up every day. Nicolas Nicolov and his technical team at Umbria couldn't have picked a better time to be deploying blog-crunching analysis. It was around then that a vice president at Yahoo, Jeff Weiner, awed by the phenomenon of blogs, marveled to me, "Never in the history of market research has there been a tool like this."

But one big problem reared its head that spring. As the snow began to recede from Flatiron, Nicolov and others began to see a new and hazardous specimen showing up in their results. Spam blogs, or splogs, they called them.

The purpose of splogs was to use the immense power of Google to cash in on the fast-growing field of blog advertising. Google offered a service called Adsense. If you signed up for it, Google would automatically place relevant advertisements onto your blog or Web page. If you wrote about weddings, the system would detect this and drop in ad banners, say, for flowers, gowns, and tuxedos. If a reader clicked the banner, the advertiser would pay Google a few cents, and Google would share the take with the blogger. For bloggers, it looked like a great way to bring in advertising revenue with absolutely no sales staff. Just click the box, blog energetically, and wait for the check from Google. But when I surveyed bloggers that spring and asked how they were faring, most of them complained. The checks usually weren't enough to keep them caffeinated, much less housed and fed.

Robots, it turned out, were running away with much of the money. These software programs spawned blogs by the hundreds of thousands (many of them on Google's free blog service) and engineered them to attract Google ads. These splogs then circulated with all the human blogs and elbowed aside millions of them to harvest valuable clicks. Here's how. Picture a future bride looking for a wedding blog. She types "wedding" into the query box at a blog search engine. The most recent post with that word in it appears at the top of the blog results. She clicks it. Chances are, she's disappointed to see a dog's breakfast of sentence fragments featuring the word *wedding*. The blog is utter trash. It's engineered by an automatic program not to be read, but simply to entice Google's robots to drop advertisements onto the page. Maybe the bride retreats from the splog back to the search engine and looks for a legitimate blog. Then again, maybe she clicks on one of the advertisements. *Ka-ching!* The splogger gets fifty cents, a dollar, maybe two dollars. The bride may not realize as she clicks the ad that she's the only human in a drama dominated by robots.

This is bound to happen more and more. Because our information travels by itself, untethered from our bodies, machines can forge and plagiarize human communication on a massive scale. This poses a never-ending challenge in the world of the Numerati: the better automatic systems understand us, the better they can pretend to be us.

The phenomenal growth of splogs in 2005 threatened Umbria's entire business. Suddenly, the company's market research was reflecting the views, concerns, and consumer habits of . . . androids. Who would pay for that? "If you don't take care of spam," Nicolov says, "it makes your analysis bogus." Initially, the Umbria team tried to weed out the splogs manually. But as the plague grew, they saw they needed to devote much of their research effort to spam fighting.

For a few frantic months in 2005, Umbria scientists struggled to teach their machines how to distinguish between the work of other machines and that of humans. For this, Umbria's scientists looked to geometry. This may come as a surprise to those of us who associate geometry with the pointy compasses and plastic protractors we carried around in middle school. But advanced geometry is a growing force in the expanding universe of the Numerati. From the king-sized laboratories of Google to small shops like Umbria, scientists often describe the world of data as a domain of sharp angles, colliding planes, and vectors shooting along endless paths.

Imagine a vast multidimensional space, Nicolov instructs me. Remember that each document Umbria studies has dozens of markers—the strange spellings, fonts, word choices, themes, colors, and grammar that set it apart from others. In this enormous space I'm supposed to imagine, each marker occupies its own patch of real estate. This is a universe that spans the quirks, the table of contents, even the punctuation of the blogosphere. Picture the theme "iPod" somewhere near Pluto, and the emoticon : (in the vicinity of the North Star. Thousands of these markers are scattered about. And each document—blog or splog—is given an assignment: it must produce a line—or vector—that intersects with each and every one of its own markers in the entire universe. It's a little like those grade-school exercises where a child follows a series of numbers or letters with her pencil and ends up with a picture of a puppy or a Christmas tree.

But Umbria's vectors aren't nearly so simple. Nicolov tries to draw a diagram on the whiteboard. But he gives up in short order. It's impossible, because in a world of two dimensions, or even three, each of the vectors would have to squiggle madly and perform ridiculous U-turns to meet up with each of its markers. The resulting map would look more like a plate of spaghetti than the straight arrows demanded by this so-called support vector machine. Even if we can't picture it with our earthbound minds, the computer has no trouble depicting the documents—blog posts and splogs alike—as vectors. They all run neatly from one dimension through countless others and, more important, through every one of their distinguishing markers. Intergalactic arrows galore. But there's a certain order to them. Documents that resemble each other, naturally enough, are neighbors in this vector space. The ones about Iraq congregate around one constellation, those about deodorants around another. A blog about deodorants in Iraq (believe me, they're out there) spans the two constellations. Blogs that have a lot in common tend to point at similar angles.

In an ideal world, the splogs' vectors would all reside in the same underworld. Then Nicolov and his team could sequester them. But in the beginning, they usually mingle with legitimate blogs. Their authors, after all, take great pains to make them fit in. This means that the Umbria team must dig for more variables, more qualities in a blog that set the humans apart. The process is similar to the fraud detection that humans have been engaged in throughout history. I remember reading about German spies in World War II who spoke perfect American English. They knew about Franklin Roosevelt's fireside chats and Betty Grable's famous legs. They could talk about high school life outside St. Louis, reminisce about dancing to Glenn Miller's big trombone at the prom. Suspicious American interrogators had to look for precise markers to set them apart. They'd ask the spies about spitballs and double plays. Maybe they tried out some old "Knock, knock" jokes. In this same way, Nicolov's team searches for variables that will betray the splogs—and set their nasty vectors careening into a neighborhood of their own.

What next? The splog neighborhood must be cordoned off, condemned. Imagine placing a big shield between the good and bad vectors. Speaking geometrically, the shield is a plane. The spam fighters maneuver it with a mouse, up and down, this way and that. The plane defines the border between the two worlds, and as the scientists position it, the machine churns through thousands of rules and statistics that divide legitimate blogs from spam.

ONE FRIGID MORNING before dawn I sit in the only open café I can find in Boulder, and I blog. It's a post kvetching about annoying advertisements on the U.S. Airways flight from Newark to Denver. Before I've finished my coffee, that post is rocketing to a computer server in New York City and up onto the blog. Like millions of other posts every hour, it sends out pings. These are updates to the computers monitoring the blog world. Thanks to these pings, search engines and blog analysts like Umbria don't have to go hunting and gathering in the world of blogs. That would take too long for a medium that changes by the minute. They simply open their digital doors and the posts arrive. It's as if they have subscriptions. Within minutes, Umbria's spam-fighting system has received my humble post and drawn it as a vector—hopefully, one flying on the safe side of the spam plane. Later it is classified by gender and generation and sentiment.

Okay. Umbria can turn my post into a vector. But can Nicolov and his colleagues turn me into a vector? After all, if each blog post can be defined geometrically, each blogger can be as well. It's just a matter of breaking down our blogs into pieces, or variables. It might analyze the subjects and blogs we write about, where we come from, the language we write in. I ask Nicolov if it's possible. Of course, he says. But it's much simpler, for now, to ignore the individuals and focus on their opinions. Each post, in that sense, participates in surveys. If U.S. Airways hires Umbria, they will see that at least one blogger, apparently a middle-aged male, has negative sentiments about their onboard advertising. This analysis is a bit like an election or the census. All voices are equal.

But as the computers cast an ever-wider net for our words and online gestures, growing numbers of Numerati will be training their vector machines on the individual. BuzzMetrics already models 2,000 of the most popular bloggers. Each is represented as an amalgam of their language, the themes they cover, and the other blogs they link to. Each of these bloggers is a hub of activity. Analysts can measure their influence and map the constellations of smaller bloggers deployed around them. With this information in hand, advertisers can buy space on targeted blogs and measure, hour by hour, the buzz that each one produces.

Can the automatic technologies that parse the words of bloggers do the same thing for some mastermind in Islamabad or London who's deploying battalions of suicide bombers? Can that person's vector be isolated like a splog? And what about others of us—you, me, Tears of Lust—whose vectors just happen to be passing through the same neighborhood? Umbria can screw up hundreds of times each day. Agencies tracking down terrorists won't have that luxury.

CHAPTER 5

Terrorist

A SCHOOL BUS pulls up beside my car. Kids stream out and make their noisy way into the National Cryptologic Museum in Fort Meade, Maryland—my destination as well. I'm a little early. Just across a broad avenue, beyond an imposing high-tech fence and a vast acreage of parking space, stands the country's headquarters of electronic espionage, the National Security Agency. I recognize the black, glass-walled cubes of the NSA from a refrigerator magnet a friend gave me a couple of years ago. It shows a bolt of lightning shooting down through the purple evening sky onto the taller of the two buildings. It appears either to be smiting the secretive workshop or imbuing it with righteous force from above, depending on your perspective. I'm here for a talk with the NSA's chief mathematician, James Schatz. It's apparently a lot easier for him to cross the street to this little museum than for me to get security clearance and make my way into the NSA fortress.

The NSA was at the center of the information war on terrorism long before 9/11. But the profile of the secretive spy agency rose following the attacks. It became all too clear that the United States lacked on-the-ground intelligence in the war against Al Qaeda. Most spies and special forces in the Mideast struggled even to make a phone call in Arabic. Few could hope to infiltrate the terror network, much less locate and capture Osama bin Laden. The answer to this shortage, for many, was to comb through digital data. "It will be their sons against our silicon," wrote Peter Huber and Fred Mills of ICX Technology, a high-tech surveillance company, in the winter of 2002.

What sorts of data would fuel the hunt for terrorists? Practically anything the government could get its hands on. In the years following 9/11, the government spent more than $1 billion to merge its enormous databases, including those of the FBI and the CIA. This would give data miners a single unified resource. But that wasn't all. They would also trawl oceans of consumer and demographic details, airline records and hotel receipts, along with videos, photos, and millions of hours of international phone and Internet traffic harvested by the NSA. This trove matched anything that the Web giants Yahoo and Google were grappling with. In May 2006, news surfaced that the NSA was secretly extending its nets even further. *USA Today* reported that major phone companies had delivered hundreds of billions of phone records to the government. These provided details on who was calling whom, from where, for how long, and whether the call was forwarded. Were the NSA staff also listening in on the calls and reading the e-mails? There was no telling. But the Bush administration made clear that when it came to antiterrorism efforts, few legalities involving congressional disclosure or court approval should get in the way. Consequently, the details of our lives flow into those databases, and it's up to government data miners to weed out the terrorists among us.

Can the Numerati at the NSA use the statistical techniques we've seen in politics and advertising to trace the path of a terrorist? Are the behavior patterns of suicide bombers similar in meaningful ways to those of foreign-movie buffs on Netflix, social butterflies on Facebook, or outlier Republicans in Greenwich Village? These are the questions I mull as I sit outside the Cryptologic Museum.

I was scheduled to have this meeting months ago. But the NSA got caught up in the controversy surrounding wiretapping done without a legal warrant, and I got put on hold. Every time I encountered one of the Numerati while the interview was pending, I asked about the challenges facing the data miners at the NSA. What I learned was sobering. The hazards of tracking terrorists along paths of electronic data are formidable, and the risks of screwing up enormous. No prudent rulers, I'm convinced, would dream of entrusting citizens' lives to these methods—unless the safety of their country hinged on it and they had precious few other options. This, sadly, appears to be the fear. And many of us may find ourselves caught in their net.

When it comes to data mining, potential terrorists differ from, say, caviar buyers at Safeway in three crucial ways. First, there's a lack of historical record. It's nearly impossible to build a predictive model of rare or unprecedented events, such as the attacks on Spanish trains, a nightclub in Bali, and the World Trade Center. This is because math-based predictions rely on patterns of past behavior. Let's say I fly to Taiwan tomorrow and purchase 200 Michelin tires with my credit card. Within minutes, MasterCard will be calling my house in New Jersey, asking if that's really me on an Asian spree. My buying patterns and those of card thieves are etched into their system. A computer program known as a neural network races through millions of transactions and establishes the limits of normal behavior. It throws up a red flag when it sees a deviation that could signal a stolen card. (This type of program detected financial irregularities on the part of New York governor Eliot Spitzer in 2007. The trail of these monies led to the discovery of payments for prostitutes and his resignation in March 2008.) But such tools are useless when it comes to recognizing or predicting something never seen before—the unexpected earth-shaking events that the author Nassim Nicholas Taleb discusses in his book *The Black Swan*.

The second problem is that suspected terrorists, unlike most shoppers or voters, take measures to blur the data signal to cover their tracks. The simplest way is to conduct important business off the network—to hold meetings face to face and send coded messages on paper or committed to the memory of human couriers. But terrorists can also manipulate the data that gets picked up, distorting what industry insiders call the "feedback loop." They can run preparations for a big bombing or hijacking, for example, elicit a response from Western intelligence agencies—and then simply hold off the attack. Jerry Friedman, a statistics professor at Stanford, compares the effect of this tactic to car alarms that go off constantly, lulling people into ignoring them. From the data miners' perspective, the non-event appears to be a false positive. They might conclude incorrectly that their algorithms need an overhaul. By tinkering with the data, the terrorists play with their heads.

Finally, failure in this realm of data mining can destroy lives. Remember Ted Kremer's shrug when his automatic reader at Umbria misread a blog post and concluded that the writer was down on Apple? Who cared? The machine got it right most of the time. Nor did it matter if Josh Gotbaum's algorithms misclassified me as a Barn Raiser or a Right Click. My pile of junk mail would be only slightly less relevant than usual. The ideal industries for the Numerati are those in which they can goof up regularly and still top the status quo. This is hardly the case in the war against terrorism. Innocent people who get swept up in the terror net can find themselves living a nightmare. This is all the more true where traditional protections, such as presumption of innocence and the right of habeas corpus, are not guaranteed, and torture is tolerated.

In the realm of counterterrorism, hundreds of millions of us are reduced to the role of supporting players, extras. The focus is no longer on us, as it was at the office and at the supermarket. The Numerati at the NSA and similar agencies around the world are attempting to track down only the tiny fraction of killers in our midst. But here's the rub. For the researchers to pick out these outliers, they must first figure out what's "normal." Picture our society on a big piece of posterboard. At first glance it looks entirely blue, monochromatic. But step closer, and you'll see tiny dots and strings of red. That background of blue represents boring, law-abiding (for the most part) us. Our only function on this display is to bring into relief the bits of red. Those are the suspected terrorists. Analysts paint that blue with the details of our lives. For this to happen, we must become known. And sometimes, if the algorithms are a little off or our behavior falls out of step with the pack, our blue gives off the slightest rosy glow. That's when trouble calls.

At precisely ten o'clock, James Schatz arrives at the museum with his press representative. He's bald and neatly dressed in a pressed white shirt and a tie. He walks with the posture and precision of a geometer. I've been briefed that policy questions will be off limits. The discussion will center on the mathematical and statistical approaches to intelligence. As we make our way into a small, sparse conference room at

the museum and set up our recorders, I recall a recent conversation with Prabhakar Raghavan, the chief of research at Yahoo. He was describing how some analysts get so tangled up
in huge amounts of data that they slot two people who should
be in the same bucket into different ones. Maybe one of them
is 51 years old, for example, and the other's 49. That bit of
data sends them into separate buckets, though there's no good
reason that it should. I wonder if there's a similar problem at
the NSA.

Schatz listens patiently as I repackage Raghavan's concerns
in a question. Are there times, I ask, when you just have too
much data? When it gets in the way and confuses things? He
seems taken aback by this line of questioning. "More data is
always better," he says.

He explains that some organizations have trouble managing loads of data. Some ask the wrong questions. Of course,
he won't tell me about the nature of data that his team is sifting through across the street. He'll only say that the "statisticians are enjoying a field day" and that "the information age
has given math a whole new life." But clearly, while politicians
and civil libertarians wrangle over how many of our personal
details should be thrown into this national security analysis,
problem-solving mathematicians at the NSA are always happy
to get more.

"The information age that we're in is going to be an entirely new era of what would be called applied mathematics,"
Schatz says. His Numerati use every statistical and mathematical tool in their arsenal—"topology, abstract algebra, differen-
tial equations, number theory"—to piece together networks,
predict migrations, analyze voices, and match photographed
faces to others in a database. Schatz says the agency has seen
"an explosion of mathematics into new areas." He describes

multidisciplinary teams, where numbers people work closely with engineers and computer scientists. The "customers" for these teams are the intelligence agents—often liberal arts grads, he says (as I nod approvingly). Presumably, they have the on-the-ground knowledge to tweak the data miners' algorithms and steer investigators toward known hotbeds of terror and menacing clans. If the agents have leads or, better, concrete intelligence, they can turn a foraging expedition into a more targeted hunt.

This small cryptology museum is a monument to the code-breaking heritage of the NSA. Governments and armies throughout history have relied on their cleverest geeks to devise secret codes to protect their vital messages. They've also counted on them to hack the secrets coming from the other side. In one of the glass cases here is Nazi Germany's famous Enigma machine, whose code was broken by ingenious British mathematicians. This was a key to victory in World War II. With the founding of the NSA, in 1952, the U.S. government built code breaking into an entire bureaucracy. It quickly grew into the largest math shop in the world (which it remains to this day, though the agency never divulges the numbers). These code breakers battled on a key front in the Cold War. While the CIA operatives met secretly with sources, from safe houses in Berlin and Moscow to thatched huts along the Mekong Delta, their counterparts at the NSA led quieter lives. They commuted to offices, first in Washington, then later to these glass cubes in suburban Maryland. Their job, quite simply, was to match mathematical wits with their counterparts in the Soviet Union.

When James Schatz signed on at the NSA in 1979, after getting his doctorate in math from Syracuse University, he stepped right into this tradition. For his first 15 years, he tells me, he worked on cryptological math, diving into some of the deepest and as yet unsolved conundrums in mathematics, and fashioning them into a sort of numeric armor to wrap around secret communications. To pierce this armor, the other side had to master some extremely heady math. Throughout the Cold War, cryptology was at the center of a mathematical arms race.

Back then, the NSA team didn't need to delve into the human psyche. Their more sociable colleagues, the spies and the diplomats, handled that murky domain. However, by the time Schatz was promoted to head the math department, in 1994, changes were afoot. The Berlin Wall was in pieces, and the United States' new enemies, whether warlords, terrorists, or international money launderers, were scattered all over the world. The challenge at the NSA was less to crack their communications than to find them. How were they organized? Where did they get their financing? What were their plans? This information didn't travel highly encrypted on secure networks. Much of it was mingling freely with the rest of the globe's chatter. Just like the rest of us, many of these villains were migrating the details of their lives and their missions onto mobile phones and the Internet. They were camouflaged by networked humanity. This meant that the mathematicians at the NSA faced a new and growing challenge. Many had to shift their focus from pure math to the hurly-burly bundles of words and pictures and smiley faces and mouse clicks that were pouring through the networks. Somewhere in that growing mass of unstructured data, they had to find the bad guys and piece together their networks. "Look at the whole telecommunications industry, all the information that's flying around on the Internet," Schatz says. "How are we actually going to tap all that for the good of mankind?" To carry out their central mission — protecting us — the mathematicians at the NSA, like the Numerati elsewhere, had to figure out humans.

THE YEAR WAS 2002. NATO troops had stormed through Afghanistan. The U.S. government was threatening to attack Iraq. And Jeff Jonas, like many others, was still obsessed with the attacks that triggered these wars. Jonas, a software entrepreneur in Las Vegas, couldn't stop thinking about 9/11. Given the information that the government possessed during the weeks and months before the tragedy, could an extremely smart data sleuth with the right tools possibly have unraveled the brewing plot and foiled it? Jonas was no expert on international terrorism, or on Islamist jihad. At this point in his life, he had never even traveled outside the United States. But he was a leading expert on finding people who want to stay hidden. He thought his approach was worth considering.

Jonas, who's now a chief scientist at IBM, is telling me this story over Chinese food at a strip mall near the Las Vegas airport. He's a catlike figure dressed in black. He leans across the table toward me as he talks, his neatly trimmed goatee hovering over my fried fish. Following the attacks, he says, he pored over public records, from newspaper articles to grand jury testimony. He was looking for paths that could (and should) have drawn investigators to the bombers. He found that two of the terrorists, Nawaf Alhazmi and Khalid Almihdhar, had been placed on the State Department Watch List two weeks after President Bush got word about planned Al Qaeda attacks. It seems easy, with hindsight, to say that investigators should have tracked them down. But Jonas notes that these two men were linked to previous attacks on the USS *Cole* and the U.S. Embassy in Nairobi. They were already targets of the highest priority. "We're not talking about people on illegal visas," he says. "That number is in the millions. We're talking about known terrorist killers in the United States. That's a small list."

If investigators had looked for them, Jonas discovered, they could have found them in the San Diego phone book. Days after landing on the watch list, the two men reserved plane tickets in their own names. Even without knowing that those aircraft would turn into bombs, investigators should have seen those names pop up. "These guys were hiding in plain sight," Jonas says. He walks through the evidence, step by step. Roommates shared phone numbers and other connections linking them to the other participants in the plot. It's true that investigators, given these details, would still have remained clueless about what this network was ultimately up to. They would have seen only that a group of people with links to past acts of terror were busy renting hotel rooms, making phone calls, and buying plane tickets. The arrests would be based on the people's records and contacts—not what they were planning to do. But detaining them may have foiled their plot. The subtext of Jonas's argument, of course, is that investigators could have located these killers by making better use of the data and the tools they had at hand.

Why have I flown all the way out to Las Vegas to spend time with Jeff Jonas? I want to find out how a society can monitor itself yet remain free and uninhibited, even sinful, by using the tools of the Numerati. Jonas is the ideal guide for this. He's vehemently opposed to the use of statistical data mining to predict the next terrorist attacks. He fears its intrusions and false alerts. Yet he trusts that data and surveillance can protect our freedoms without sacrificing our privacy. His

method isn't so different from that of an old-fashioned detective. It starts with a lead: a suspect, a door to knock on, a sign
of suspicious behavior. And it follows the trails of data from
there. This is what I call the gumshoe approach, a focused
alternative to predictive data mining. Jonas told me months
ago, over lunch in New York, that Las Vegas was a perfect test
case for gumshoes. That's what I've come to see.

Jonas has built his business and his fortune on following
threads of data. He began developing his targeted approach
by tracking a group of aquatic killers and their victims. It was
early 1995 when the young software whiz first arrived to work
at the Mirage Hotel in Las Vegas. The fish swimming in the
massive aquarium at the Mirage were worth $1 million. But
there was a problem. Fancy fish were disappearing—presumably into the mouths of their tank mates. Jonas's job was to
build a tracking system for fish so that the casino could calculate the survival rate for each type—and avoid investing in
Darwinian losers. Jonas, a supremely social animal, learned a
lot about the casino business while acquainting himself with
the movements of tropical fish. Business was booming, he
heard, and this was creating vulnerabilities. As thousands of
people streamed through their doors, the casinos were finding
it harder to keep their eye out for thieves and cheaters. They
needed a system more advanced than the one Jonas was building for the aquarium. In the hunt for humans, they would
be looking for specific predators. For decades, the casino had
entrusted this work to people. But things were getting out of
hand. The numbers were too big.

So Jonas built software to help casinos pinpoint known
cheaters, grifters, and goons—what casino execs refer to as
"subjects of interest." Called NORA, which stands for non-ob-
vious relationship awareness, it specialized in rifling through

many different pools of internal casino data, from personnel files to credit applications, looking for common threads. NORA might see, for example, that Krista, who was on the suspect list, had the same home phone number as Tammy, who had just applied for a job as a blackjack dealer. Were they partners in crime? NORA highlighted the correlations. Then it was up to humans to dig for the answers. But piecing together these connections from a sea of data was invaluable. NORA, quite simply, helped straighten out who was who.

NORA, unlike other data-combing setups, doesn't look just backward in time. It also reaches into the future. Let's say a casino is on the lookout for the leader of a gang who uses the Internet to put together teams of crooks. (This is a growing problem.) They've dug around and have a few facts on him—say, an alias, two phone numbers, and an address. In a traditional backward-looking data hunt, investigators comb through all of their records looking for signs of him. Nothing turns up? Thank goodness . . . But what if the very next day a friendly, big-tipping tourist checks in at the casino. The phone number he scrawls on the registry form is one of the two from the list. In a traditional system, the casino won't spot him until they hunt again. But with NORA, each new piece of data —each phone number, each name and address—creates a new query. It asks the system, "Hey, anything fishy about this one?" That's how NORA extends forward in time. It's constantly at work, prospecting into the future, assembling the bits of evidence as they arrive.

After Jonas unveiled NORA, it was just a matter of time before other companies and government agencies began knocking at his door. The challenges of finding identities and tracing connections within vast databases were hardly unique to Las Vegas. Anybody interested in sorting through data to find and profile shoppers or patients or voters or workers or lovers . . . in short, entire ranks of the Numerati desperately needed NORA, or something like it. Those hungriest for NORA were wrestling with the biggest and messiest sets of data in the world as they searched for the identities and movements of terrorists. In January 2001, In-Q-Tel, the venture-financing arm of the Central Intelligence Agency, bought a share of Jonas's company, Systems Research and Development (SRD). And following the attacks of 9/11, NORA was enlisted in the war against terrorism. Four years later, IBM bought SRD for undisclosed millions. This made Jonas a rich man and turned the entrepreneur into a distinguished engineer and chief scientist at Big Blue. His start-up morphed into IBM's Entity Analytics group. In his new role, Jonas has a lot to say about the use of technology for national security. He sits on panels, testifies at inquiries convened by the president, and is a leader in IBM's efforts on the defensive front of this new war.

Jonas tells me that he was a beach bum and guitar player as a kid. But in the tenth grade he signed up for a computer class and then took another one. When he ran out of computer courses, he told himself, "I'm out of here." He passed the graduate equivalency exam and dropped out. Within two years, Jonas was running a booming software business, Preferred Programming Services. But he knew a lot more about writing software than running a business. The debt got out of hand, and he went bankrupt. By the time he was 20, Jonas was sleeping in his car.

He clawed his way back into the software business. Even before finding a place to live, he started up his next company, SRD. The business took off. But at age 24, Jonas hit another big turning point. While checking out a new BMW, he had a salesman take him for a ride—and the salesman drove the car off the road. Jonas broke his neck. For a short while, he was paralyzed. After he regained use of his limbs, he worked his way through a long rehab and back to fitness. But the center of his spinal cord, he says, is dead. And to this day he has only slight feeling on the right side of his body. "I can feel the difference between the point of a pencil and the eraser," he says, poking his fingers with a fork. He grips a glass of ice water. "But I can't tell the difference between heat and cold. And I don't feel pain very well." Ever since the accident, Jonas has been a perpetual motion machine (with a sky-high threshold for pain). He schedules his business meetings and conferences in places like Singapore, Brazil, and New Zealand to coincide with Ironman Triathlons. One day he's telling a roomful of executives how to locate people with software, and the next day he's swimming across a bay, biking up and down a mountain, and running for as long as 14 hours straight. He says he thrives on such episodes. Sometimes, he says, he peels off his shoe and sees that he's ripped off a toenail.

"I think I can get you into a crow's nest," Jonas tells me one afternoon. He's referring to the surveillance room high above the casino floor. He's eager for me to look at the world through the security managers' eyes, not so much for all the details they can see from up there, but for what they ignore. This is the key to all kinds of surveillance, he says: what to focus on. It's crucial in Las Vegas. People come here to do all kinds of things they don't dare try back home. They crave the freedom to blow money, drink way too much, and follow virtually every animal impulse that stirs them, from staring down a bartender's dress to arranging a ménage à trois—all off the record. In short, they want to sin anonymously, which is another way of saying that they're looking for freedom and privacy. For this, oddly enough, they come into a dark world bristling with cameras. It may be a visit to the future, in which cameras and other sensors surround us and protect us from harm. We can only pray that the powers that be will stick to that mandate and respect our secrets. Jeff Jonas argues that casinos have come as close as anyone to mastering this balancing act.

"THERE IT IS."

"What?"

"See what she did with her hand?"

"Play it again."

Thanks to Jonas's connections, I'm up in the crow's nest of a major casino in Las Vegas. It's dark. Most of the light comes from the dozens of TV monitors blinking from the wall. Four of us are gathered around one of these TVs. We're looking at a young woman who's drinking and joking with her friends and having the best of times—while she cheats at the blackjack table. To be fair, we're studying every hand she has played, and we see her cheat only once. It happens quickly. The cards are dealt. Her bet is placed. She has a blackjack, a winning hand. And with a lightning gesture—which I find especially impressive, considering how heavily she's drinking—she adds another $5 chip to her bet. That's illegal, a gaming violation, as they call it in Las Vegas.

Granted, she hasn't sneaked loaded dice onto the craps table or filed down the ball bearings in the roulette wheel. But her move is a crime, and I'm privy to the discussion in this digital crow's nest about how justice is to be administered. This is life at the heart of a surveillance society, I think, as I look around. It's an impressive view. The casinos all have fixed cameras staring down at every table. Tracking cameras cover nearly every inch of floor space. All of these send video to the banks of TV screens in front of us. When I arrived, we test-drove these tools by following one customer, a man who had just walked into the casino. He was carrying a backpack over one shoulder. His eyes were probably still adjusting from the desert glare outside to the dusky shades of the gambling den. The surveillance team tracked his movements, switching from camera to camera. They exchanged terse coordinates, like pilots on a bombing run. We watched from screen to screen as the man made his way through the banks of slot machines, past the craps tables and the bar, and finally to the hotel's front desk. I was hoping, for his sake, that he wouldn't scratch himself or stick an idle finger into his ear along the way. He was performing for a crowd.

A few minutes later, the surveillance team got word from the floor to study the behavior of a woman at a blackjack table. She's the one we're watching now. She's in her twenties, I'd say. She's playing with two others, both men. They're smiling and joking. She's wearing a low-cut blouse with spaghetti straps. She takes sips from her drink, which she holds with her left hand, and she occasionally reaches up to fix a strap that keeps falling from her shoulder. On one monitor we watch the hands she's playing. She has two cards totaling 14 and asks for another. She gets a king. Whoops. That puts her over 21, which means she's lost the hand. It hardly seems to ruin her fun. Meanwhile, at another monitor, someone else on the team has rewound the tape. (Yes, it's a VCR, which seems surprisingly primitive to me.) He's watching every hand she has played. He's the one who sees her cheat. "Got it," he says, and promptly cues it up for us. We watch the illegal move again and again in slow motion. Isn't the dealer looking right at her? Didn't he see it? Did the others do something to distract him? It's hard to tell.

Still, the casino has concrete evidence that she has broken the law. A discussion ensues. Is she a pro? Is she drunk? Is it possible she doesn't know the rules? They decide she's no pro. It's only $5, after all. They send a supervisor on the floor to talk to her as she and her friends are leaving the table. We watch. She's surprised, confused, and then grave. Then the supervisor says something that puts her at ease. She relaxes, smiles, jokes, and then goes her tipsy way. The authorities have let her know that she, along with all the other gamblers, lays her bets under a legion of watchful eyes. But they've opted this time not to dampen the partying mood. In fact, as she walks away, they still don't know her name. It doesn't matter.

The same goes for the other gamers milling about below. Whether they're sidled up to the bar downing Cuba Libres or grimly feeding quarters from a paper cup into a slot machine, they're free. No finger wagging here. If they pay with cash and look older than 21, no one asks their names. They're anonymous. And yet they're at play under the gaze of pit captains and security forces on the floor and their teammates up here in the surveillance room.

What details, I ask the boss in the crow's nest, does the casino need to collect in order to pick out the handful of grifters and thieves? What data separates them from the rest of us? He says it boils down to three questions: Are they on the casino's list of known crooks, cheaters, and card counters? Does their behavior in the casino signal malicious designs? And are they winning lots of money? If you think about it, these three questions are the underpinnings of most police and intelligence work: Does the person have a record? Is he acting suspiciously, perhaps in cahoots with others? And is he in or near the spot when significant events take place, whether they're bus bombings in London or a fabulous pinch-me-and-tell-me-I'm-not-dreaming run on a craps table in Vegas?

It's the folks here in the crow's nest and on the floor who scrutinize the behavior of the gamblers. The signals they're looking for are far too sophisticated and nuanced for machines or data mining to pick up. Some people, for example, aren't smiling or drinking. "They look like they're working," he says. Some of them make gestures that could be signals. They rake their hands through their hair insistently, or they make a tipping gesture with a drink. Some put tiny card-counting computers into their shoes and make small movements as they hit buttons with their toes. These gestures fit into patterns surveillance teams are taught. Picking them up requires observation and human intelligence.

The far easier signals come from numbers. They should be just as predictable, on average, as the trains pulling into a Zurich station. In the short run, they fluctuate, giving hope to long-shot gamblers. But with time, each tool hits its standard rate of return—each one of them favoring the casino. When casinos see deviations from the expected numbers, they go take a look.

The third category comes from data. That's where things have changed—and where people like Jeff Jonas make a difference. In the early years, Las Vegas relied on plugged-in people to collect data. Experts combed through hotel, credit, and personnel files, looking for folks who appeared as "subjects of interest." They zeroed in, of course, on known crooks and cheaters who should be shown the door—or locked up. But the data hounds were also on the lookout for high rollers who should be upgraded to luxury suites or comped a magnum of champagne. When it came to spotting something out of the ordinary, whether promising or suspicious, no machine could rival a smart and experienced human. Think of Humphrey Bogart's character Rick in the movie *Casablanca*. He'd eyeball the floor and scan the ledger names. He knew the sordid stories. He kept up with the evolving tangle of friendships and alliances. He had the place scoped out. But these jumbo casinos have grown far too big for the human approach. Some of them have more than 3,000 rooms. They entertain 100,000 visitors in a single day—considerably more than Rick's gin joint in West Africa. They need powerful machines to sort through the data. And the approach the casinos use, Jonas says, could help sharpen the focus in the battle against terrorism.

Of course, we can't all punch the pillow and roll over, confident that the most dangerous terrorists will pop up on government watch lists, obediently publish their names and addresses in phone books, and reserve plane tickets and hotel rooms for their suicidal cohorts. That would be too much to wish for. We don't have all the suspects on record, not by a long shot. The gumshoes are short on leads. Data miners, meanwhile, often struggle to find meaningful signals. Does this mean that the government calls off the electronic hunt for terrorists and the Numerati retreat to safer and surer jobs in advertising and grocery stores, where their statistical methods work fine? Not on your life. The need to close the intelligence gap is urgent. Our safety is at stake. So we reach for something, even if it's faulty, to protect us. In cultures that are tough to penetrate and understand, data mining at least offers the possibility of finding something. In essence, we compensate for our shortcomings in languages and on-the-ground intelligence with a heavy dose of unproven technology.

This is fueling a researchers' gold rush, a period of wide and wild experimentation. The Numerati are reaching into any discipline they can find, whether it's economics, physics, biology, or sociology, to unearth formulas that can be tweaked to predict the behavior of terrorists. They're not only in the business of mining data; they're also mining theories, many of them hatched long before computers were around to crunch the numbers. The idea is that certain patterns, both in human behavior and in nature, pop up in different realms. Maybe some of them will help expose sleeper cells or bomb factories. Researchers have centuries of data, for example, on the diffusion of plagues and epidemics. They can tell you, mathematically, the chances that the seeds from my dandelion-infested yard will float onto my neighbor's pristine, manicured lawn across the street. Do the hateful ideas of terrorists spread in similar patterns? Do terror cells metastasize like cancer? Do they mutate and evolve like certain viruses? No? How about if you change a variable or two? Social scientists study the evolution of networks, from those on MySpace to cell phone users in Singapore. Who are the hubs in these networks? How do they rise to this status? Do their spheres of influence shift with time? Again, what researchers learn here can be boiled down to the mathematics of human communication and organization across networks. Does Al Qaeda follow similar patterns?

JUST AS OUR experience on the Internet is moving beyond the written word, so is the data pouring into NSA computers. Much of it arrives as spoken voices, images, and video. It might be a face in a crowd in Baghdad, or perhaps a raspy voice giving orders in Farsi from a Skype account somewhere in the Horn of Africa. To mine this outpouring of data, machines must make sense of the words we speak in scores of languages. They must learn to pick out one or two faces from six billion others. To extend their nets into sounds and images, the counterterrorists need new technologies. Researchers around the world, many of them scooping up rich government grants, are busy assembling them.

This technology has been replayed in the movies so long that it seems familiar. A machine automatically sorts through photos of people at a café in Tripoli or Karachi, or perhaps crowd shots at the Olympics. Then it matches the faces in the photos with a dossier of known and suspected terrorists. That's the goal. As these systems take shape, our faces will find their way into enormous databases. Then computers run by governments and corporations will be able to map the movements of humanity. For most of us, in truth, this won't make much difference. Our faces will show up along the same trails drawn by our airplane tickets, credit card bills, and—above all—our cell phones. Yet these facial images could prove vital for police. They could capture data on people who are struggling mightily to stay off the grid. A photo reader might find, for example, that the same green-eyed man with a bump in his nose and a scar on his lip has traveled at least three times this year between Newark, the rough Parisian suburb of Saint-Denis, and Cairo. Does that face pop up on other databases?

A global snooping network is already emerging. Britain has been an early leader in installing security cameras, with 200,000 operating in London alone. The image of the average Briton, police say, is captured by as many as 300 cameras per day. American cities, including Chicago and New York, are rushing to follow suit. And late in 2007, according to the *New York Times,* the Chinese government announced plans not only to monitor the streets of the southern city of Shen-zhen with 20,000 police cameras but also to give police there access to the feeds from another 180,000 video cameras run by the government and private companies.

All of us, from bombers to subway passengers, will be playing ever bigger roles in these surveillance films. But on this global stage—unlike the cozy casinos in Las Vegas—there aren't nearly enough human workers to monitor all the action. And the machinery to sift through all this video isn't yet up to the job. At this point, an automated system can com-pare mug shots of suspects with thousands of photos on file, and suggest a handful of them that have a similar facial profile —before handing over the job to humans. Despite what Hol-lywood would have you believe, identifying faces in the real world is still very much a work in progress. Faces duck in and out of shadows. They turn from full face to profile. They tighten as we laugh and bulge as we eat. With age, they sprout beards, lose teeth, gain heft, grow new lines and furrows. Pin-pointing the same face through all those changes is an im-mensely complicated task for a machine. But the computers are getting closer. The U.S. National Institute for Standards and Technology held a Grand Challenge for face-recognition systems in 2006. Researchers had to develop 3-D models of faces so that they could be recognized from a variety of angles. In the four years since the previous competition, results im-proved by a factor of ten.

Scientists are also delving more deeply into the noises we make. They're analyzing not just the words we utter but even the timbre on our voice. Researchers at BBN near Boston, for example, have government contracts to study the effects of emotion on our voices. "When you're under stress, you'll pro-duce sounds differently," says Herb Gish, the chief scientist at the company. "Are these different from when you're angry, or sad?" Naturally, they tackle this challenge, like so many others, by breaking down the voice into bits of data. They study the patterns, almost as if they were strands of DNA, and correlate them mathematically to the emotions they express. At some point, Gish says, researchers will have tools to gauge the likelihood that the voice echoing through the phone line or over the Internet comes from a person who is sad, angry, or tense. This means more work for data miners. They'll have to write algorithms to burrow through vast audio files, searching not just for key words or network patterns but certain moods. The level of complexity shoots ever upward.

As government sleuths dig deeper into networks and data, they find themselves wrestling with the same challenges as the Numerati elsewhere, at Google, at Umbria, and at Microsoft. Spies and advertisers are working on the same math. This leads to a battle for precious brainpower. A generation ago, the NSA could lay claim, in its quiet way, to many of the brightest mathematicians and computer scientists in the country. But now they have to compete with Internet giants worth hundreds of billions of dollars. There's a global race for star talent. When hotshots like Yahoo's Raghavan, a former star at IBM Research, step into the Internet job market, bidding battles erupt. In Raghavan's case, it was between Microsoft and Yahoo. Both companies struggle to keep up with Google, which has been minting millionaires all over the world. How can the NSA compete? What's more, the Internet companies are free to open research divisions in India, China, Japan, and Europe—which produce more mathematicians and scientists than the United States does. And as we've seen in nearly every chapter of this book, these companies hire plenty of gifted foreigners in the United States. The NSA, by contrast, is limited to U.S. citizens—a severe constraint. Schatz says the agency can still land great ones, people who are drawn to a settled suburban life, national service, and a chance to grapple with outsized challenges.

Still, when it comes to tracking down the likes of Al Qaeda through data, the government can hardly do the work alone. So they farm much of it out. They take vast files of so-called terrorist data. They declassify the files by changing names and other features. Then they distribute these sets to university and corporate researchers. This opens the job to thousands who find themselves, as I do, outside the fenced compound of the NSA.

HAVE YOU EVER met someone and thought, "Isn't it strange that I didn't meet this person before?" It's usually someone whose path crosses with yours. Maybe you live in the same neighborhood. Maybe you ride the same train every morning, or you have an ex or two in common. Perhaps you are the only two in your sleepy town who have pierced tongues and green hair.

Now imagine trying to predict the next friend you'll meet. Which of your circles will this person emerge from? Which facts about yourself, and about others, are most likely to lead to the connection? Researchers at Carnegie Mellon University are searching for answers as they wade through gobs of unclassified surveillance data from the Department of Homeland Security.

Let's say that three suspected bombers were spotted a week ago in Nairobi. There's no sign of them now. But chances are, they were making plans with comrades, perhaps members of sleeper cells. Who are these hidden allies? Often, says Artur

Dubrawski, one of the CMU researchers, the government has data on lots of people. It knows that a few of them are suspected terrorists, but everyone else is just a name. So how do investigators know where to look? Which of those names are most likely to be associated with those three who passed through Nairobi?

It's easier to imagine the math involved here by picturing your own life and your own friends. Say you're having four people to dinner, and you want to find a fifth guest. This would ideally be someone who would fit neatly into the group, either through shared friendships or values. This is what the CMU team would call "a next friend." You and the four people in this tiny example are going to be the training set. You're going to add up the various links you share with those people and then use them to calculate the most likely fifth guest. So go at it. What features do you share with these people? Let's say two of them are lawyers, like you. Three are women. One is a friend of your sister's. One is a former lover. One of the two lawyers went to summer camp with you back in the 1980s. He lives upstairs. Two of them speak excellent French, a language you love, and a third cooks good French food. Some of the features may seem outlandish or beside the point. Maybe you know for a fact that three of these people snore, or that two of them used to date diplomats. Include everything. The irrelevant stuff—the noise—will be flushed out later.

Now picture yourself and your four guests as five dots on a graph. In the world of social networks, these are called nodes. (And most training sets have hundreds or thousands of them.) Each link shared among the five people is a line connecting them, a so-called edge. In the computer world, these graphs exist in limitless dimensions, like Umbria's blog universe. You don't have to worry about these thousands of edges running into each other and making a mess, the way they might in a grade-school science project. There's space for all of them. The next step is to calculate the importance of each edge. This involves statistics. Which links most distinguish your friends from everyone else in the world? If you're inviting both men and women to your dinner party, the gender links are meaningless predictors. In that respect, your party reflects the world. The Next Friend approach zeros in on the links that set your party apart. The lawyer links, for example, the French, the summer camp connection, those are probably much more predictive. So they're given a higher score, or coefficient. Those lines on the graph are thicker. Now it's time for CMU's program to put all the numbers together and create a composite profile of your most likely "next friend." Then it will go through a database of your friends and give them each a score. The highest-ranking person on the list is its choice as the most likely to enjoy dinner next Saturday with you. How did that person get such a sky-high score? In this case, it might be a taste for French food, some experience at summer camp, or perhaps a litigious nature. In the more important national security case, the next friend of the three terrorists seen in Nairobi might turn out to have been in Afghanistan during the same period as two of the others. Or maybe they call the same phone numbers or have a brother in the same jail.

That's assuming that the declassified database in CMU's computer is bursting with the same kind of rich details that you came up with to describe your dinner guests. This brings us to a central problem in the electronic hunt for terrorists: iffy data and incomplete files. It shouldn't come as any surprise that we know our friends far better than we know our enemies.

Intelligence services are often flummoxed by even the most basic piece of data in a person's file: his or her name. This is one crucial area where our cultural diversity defies the sorting and counting magic of the computer. Jack Hermansen knows this all too well. He's been working on the electronic recognition of names since 1984, when he got his doctorate in computational linguistics from Georgetown University. The U.S. State Department called on him back then to help figure out which names belonged to which people. It seemed like a simple enough task. Figure out the variations, from culture to culture, on the spelling of a name like Sean, Mohammed, or Chang, and stick them into a computer. "They wanted some linguistic fairy dust sprinkled on their problem," he says. But Hermansen knew that the global interpretation of names was endlessly complex. That "Haj" in an Arab name? That just means he's made the pilgrimage to Mecca, but it will show up as a last name in some databases. A Chang will appear in French as Tchang, or maybe Tchung, and the Germans and Russians will have different takes on it. The Chinese alone have 11 different spellings for Osama bin Laden.

Hermansen built a name-identifying company, Language Analysis Systems (LAS). Its immense mosaic of names, as they're organized and spelled from one culture to another, embodies the labor of anthropologists and linguists, not computer scientists. But that expertise gets plugged into computers. In spring 2006, Hermansen sold LAS to IBM. Now it works closely with Jeff Jonas's identity detection unit. Despite progress, Hermansen says, untangling global names will continue to confound us for generations. "My grandchildren could be working on this."

This doesn't mean that techniques like the Next Friend analysis won't be useful inside the NSA. But those multidisciplinary teams Schack talks about will need lots of guidance from the intelligence officer. Hopefully, that liberal arts grad masters a foreign language or two. To understand the cultural biases in selecting certain data, a strong dose of anthropology can't hurt.

In the meantime, many tools and technologies that are being honed for national security may find receptive markets closer to home. It makes sense, doesn't it? The intelligence agencies may have spotty data in their foreign files, with all sorts of tangled and duplicate names. The files that we inhabit, by contrast, are brimming with useful and intelligible data. Our records at work, for example, have clear names and schedules, and everyone works on the same e-mail system. Consider how Next Friend analysis could come in handy. Say a colleague leaves a job at your company. Who's the person most likely to be affected by his departure? Will that person be the next to jump ship? Managers can intervene. On a darker note, what if a colleague is caught selling confidential information? Could a Next Friend analysis point to others who should be monitored? Scary perhaps, but look at the bright side: once the Numerati master these techniques on us, maybe they can use them to catch terrorists.

BACK IN LAS VEGAS, I try that same quip on Jeff Jonas. He doesn't find even a hint of humor in it. The way he sees it, the technology to monitor us, and to predict our behavior, will continue to march ahead endlessly. "Everybody is trying to compete," he says, "whether it's governments competing against each other, companies competing against each other, or governments competing against [terrorist] threats. And when you compete, you want to have the best human

resources, the best minds, the best tools, and the best data.
You always want more data, better tools, smarter people."
He leans forward. "When's that going to end?" He answers
his own question. "It's not," he says. And the machines that
gather our data will continue to proliferate, with wireless sensors and cameras following our movements. This could mean
casino-style surveillance in much of the rest of the world. "I'm
thinking, man, when is this going to slow down?" Jonas says.
"Is there any throttling mechanism? Whoa, whoa, whoa!"

The throttling mechanisms that he's talking about—locks
and shields to defend our privacy—often don't appear in the
original technology. They're tacked on later as fixes. Why so?
The nature of innovation is first to create the breakthrough
service or product. Control mechanisms, like privacy filters,
come afterward. Jonas himself, he admits, was late to grasp
the implications of his own inventions. There he was, building the identity systems that could become pillars of global
surveillance and, he says, "Honestly, four or five years ago, I
didn't even know what the word *privacy* meant!"

Here's the prime danger he sees. Imagine that government
data miners sift through the details of our lives but fail to uncover terror cells. Good chance they'll likely push for more
data—making the case that collecting it is a matter of national
security. Mishandled or misunderstood, this hunt threatens to
ransack our bedrooms and medicine cabinets, ripping away
what privacy we still hold on to. It can implicate innocent
people—"false positives" in data-mining lingo. Statistics, after
all, point only to probabilities, not to truth.

The damage can spread even further. Let's say that terror sleuths, while failing to find terrorists, come across other
interesting patterns in our data. Maybe one of us looks like
a tax cheat. Another belongs to an informal e-mail network

that includes pornographers. What then? Will privacy advocates and civil libertarians dare to defend suspected pedophiles? Or imagine that your accountant quietly runs an illegal side business as a bookie. Suddenly, says David Evans, CEO of Clairvoyance Corp., a data analysis company in Pittsburgh, "you get analyzed for any other data that support the relationship with the bookie. Where did that money come from? Were there withdrawals in cash? These are the statistics that could be used to create a case against you." This, says Jonas, is why we need technology to protect our identities and policies that safeguard our rights. "We're going to need some *smart* people in policy," he says. Without sound oversight, we're liable to get the worst of both worlds—a surveillance society that still fails to make us safe.

Jonas says he didn't understand these risks when he was building NORA. Then he began to see how once the information about us in databases is compiled, it could be used in different and devious ways. He calls it "repurposing." This late awakening has turned Jeff Jonas into a champion of privacy. He's built a strong privacy-protection supplement for NORA. He originally named it ANNA, a play on "anonymity." Now that it's gone corporate, it's been redubbed IBM Anonymous Resolution. It encrypts each identity into long series of letters and numbers known as "one-way hashes." Governments and companies can then search for connections—combing through the passenger list of a cruise ship, looking for the hashes of suspected terrorists. This system reduces the risk of data leaks. More important, no one sees the names until a match appears and the company gets a formal request to reveal the identities. With Jonas's system, our most sensitive data is exposed, but not attached to a person's identity. But even as some of the Numerati come up with schemes to protect us, others are racing ahead in the data hunt for the next would-be bombers. They're bound to learn lots about us in the process, more than most of us are ready to share. "We technologists had better spend a little more time thinking about what we're creating," Jonas says.

CHAPTER 6

Patient

I REMEMBER taking my mother, in her last year, to the doctor. One of us had to accompany her, to write down the new dosages of the eight or nine medicines she was taking. She was frail and a bit forgetful, and much too busy taking care of her blind 95-year-old husband to think much about herself. Her doctor asked questions. Does that hurt? Well, yes. And that? Um. Yes, a little. He scribbled these hazy answers in his pad.

"Are you having trouble sleeping?" he asked.

"No." She was pleased to provide a definitive answer.

I jumped in. "Mom, weren't you up making cocoa in the middle of the night?" Well, she allowed, some nights were better than others. The doctor kept scribbling. I was already researching this book, and I remember wondering, What kind of data is this?

Miserable data, is Eric Dishman's answer. From his research lab at Intel, just outside Portland, Oregon, Dishman is working feverishly to replace the fog, forgetfulness, and wishful thinking of human memory with minute-by-minute updates pouring in from electronic sensors. This 40-year-old anthropologist has a broad face, with a burr of dark hair on top and a smile stretching across the bottom. He's an evangelist for a new approach to health care because he sees the status quo as untenable. His words spill out in torrents: "I took an unpaid leave to go deal with my wife's grandmother, who fell and died from the fall," he tells me, as five members of his team and I chow on Chinese takeout in a conference room. "Actually, she died from horrendous medical errors that occurred," he says. "Then my grandfather fell. He didn't die, but we put him into a nursing home in North Carolina. I'm not trying to be egocentric, but I'm a world-known expert in this field, and I couldn't stop this from happening. I work at Intel, I make a decent living, I know the technology, I know every CEO in this industry, I can call senators on these issues." And yet people in his own family fell victim to the very health-care disasters he works to prevent. "I lose one relative, and another gets forced to move into a nursing home," he says. "What happens if you don't speak English, if you don't have the kind of access that someone like me has? It's terrifying."

He thinks that over the next generation, many of us will surround ourselves with the kinds of networked gadgets he and his team are building and testing. These machines will busy themselves with far more than measuring people's pulse and counting the pills they take, which is what today's state-of-the-art monitors can do. Dishman sees sensors eventually recording and building statistical models of almost every aspect of our behavior. They'll track our pathways in the house, the rhythm of our gait. They'll diagram our thrashing in bed and chart our nightly trips to the bathroom—perhaps keeping tabs on how much time we spend in there. Some of these gadgets will even measure the pause before we recognize a familiar voice on the phone.

A surveillance society gone haywire? Personal privacy in

tatters? Not at all, says Dishman. He predicts that many of us will deploy these sensors to spy on ourselves in order to live healthier, happier, and longer lives. We'll do it, in other words, because we choose to. And we'll come to learn about this technology, insidious as this may sound, by trying it out on the old people we love—especially those who live far away. Take it from me. As Dishman walks me through his lab, a year after the deaths of my parents (and only ten miles from their home), I look at each new sensor and think, "Boy, we could have used one of those."

Here's an example. At that checkup I went to with my mother, the doctor told her to weigh herself every day and to keep a daily log of the results. This was important, he said, because a dramatic rise in her weight might indicate that her weakening heart was failing to pump fluids. He didn't go into the details, but untreated, those fluids would fill her lungs and kill her. Later that afternoon, I bought an electronic scale. I knew, even as I showed it to her, that this plan was fatally flawed for at least three reasons: She wouldn't always remember to weigh herself. She would find it very hard to tap the scale with enough authority to activate it. And even if she succeeded, she'd have a tough time reading the electronic display, which even I squinted to see. In short, even though the doctor needed updates and my mother was willing to furnish them, crucial sensor and recording components—my mother's eyes and her memory—were not up to the job. (For my father, blind and increasingly immobile, daily weighing was nearly hopeless.)

Dishman guides me toward a small tiled section of kitchen flooring. It's a prototype, which he calls the "magic carpet." Under each of the beige tiles are webs of weight sensors. If my mother's kitchen had been equipped with a magic carpet, her every visit to the kitchen would have dispatched details on her changing weight along a wireless path from the tiles to her computer, and from there to the doctor's office. From this thread of data, the doctor would not only be able to monitor her weight, but equally important, he could receive an alert if one day she failed to walk into the kitchen. That's a fact worth knowing.

I can just imagine my parents laughing at the extravagance of a magic carpet in their kitchen. It sounds like something from the Jetsons. Yet in the past half-century, while medical costs have skyrocketed, electronics have gone the other way. Back in the 1960s, when my parents were raising us, our doctor paid affordable house calls if we had the sniffles—and NASA spent millions of dollars for computers no more powerful than the battered cell phone in my pocket. Consider how dramatically things have changed. For the price of one single bottle of my mother's heart medicine—$80—she could have bought a wireless network for their home. (Believe me, I encouraged her to. Their dial-up modem drove me batty.) She could have replaced her buggy old computer for the cost of one MRI. Near the end, my parents were spending about $180 a day for home nursing. For just a fraction of their monthly nursing bill, they could have thrown enough blinking sensors and networking gizmos into their house to record and transmit every step, bite, breath, word, and heartbeat in their Portland house.

But who would have noticed this river of data? My parents, like many of us, had a hard time getting responses to one of the simplest and most definitive alerts imaginable: a phone call to the doctor's office. If doctors are so understaffed that they struggle to return phone calls, I ask Dishman, how are they ever going to interpret data pouring in from magic carpets and countless other devices? "That's precisely the point!" he says. The doctors are too busy. The gadgets by themselves don't help much. It will be up to the Numerati to pore over the patterns of movements and speech and social interactions and then figure out what they mean. Only good math can sift through these floods of nearly meaningless data to provide doctors with specific alerts. This isn't easy. In one Oregon study, people's beds were wired to monitor their nightly movements and weight. One woman, researchers were startled to see, gained eight pounds between bedtime and breakfast. A dangerous accumulation of fluids? Time to call an ambulance? No. Her little dog had jumped on the bed and slept with her. Culling the pugs and corgis from the data will be up to the Numerati.

Even the simplest of these algorithms must be customized. For some invalids, for example, it's a red-light alert if they're out of bed. Maybe they've fallen or are teetering in the hallway or fiddling with the stove. For healthier patients, it's the failure to climb out of bed at the usual time that spells potential trouble.

This analysis is still in its infancy. Think back to the Internet in the mid-1990s. As we learned to send e-mails and call up Web pages, we were creating data. But it took a few years for data-crunching companies like Tacoda, Umbria, and Google to learn how to analyze our clicks and search queries and blog posts, and build businesses around them. Dishman's job now is to entice us to use his sensors. Only by hooking us up can he generate the streams of data that the Numerati feast on.

The gadgets won't make their way into many of our homes until they pass important tests. They must be easy to use and provide decent service, while also protecting at least a bit of our privacy. If these machines create confusion and frustration, they'll wind up stacked in a closet, gathering dust, like that digital scale I bought. And if users have reason to fear snooping, from marketers, scam artists, or insurance companies, they'll likely pull the plug. These are the challenges ahead for the electronics and software companies, such as Intel, Microsoft, and Google, that are rushing into the medical business.

Dishman sees this march of medical monitors as inevitable. Aging societies around the world face exploding healthcare costs, especially as the jumbo generation born after World War II starts retiring in droves. It creates a market for automation, which Dishman spotted long ago. He worked to develop the science in the 1990s as part of a start-up financed by Paul Allen, the cofounder of Microsoft. But the key to getting these monitors into hundreds of millions of homes was to harness the power and reach of a global computing giant. He knocked on doors throughout the tech world, making the case that home computers would become, among many other things, home nursing stations. But those companies, he says, fretted about saddling their youthful brands with geriatrics. Intel finally relented. Dishman launched the home health division with just one colleague in 2001. Two years later, they issued a press release about predicting Alzheimer's disease. The public response was immense. It was led by people just like me, Dishman says, who were avid for technology that could keep an eye on their elderly parents. Since then, his division has opened a research branch in Ireland and has carried out tests in more than 1,000 homes in 20 countries. He runs a national nonprofit for home health care, which involves 500 companies and universities.

This push to develop electronics isn't just a matter of replacing doctors and nurses with machines or using digital readings to supplement our faulty memories. Constant monitoring is bound to change the very nature of health care, eventually giving each one of us the kind of continuous health surveillance historically reserved for VIPs, such as vice presidents with heart disease, billionaires, and astronauts. This change, says Dishman, shifts the focus from after to before, from crisis response to prevention. If the Numerati get it right, they'll note changes in our patterns of behavior long before we fall ill. They'll know our typical daily routines when in health. Then, when they detect changes in our activities, they'll figure out what we're coming down with and start treating our maladies before we get them — or at least before we perceive them.

In many ways, these promises of preventive care echo others coming from genomics laboratories, another growing empire for the Numerati. In universities and pharmaceutical labs around the world, computer scientists and computational biologists are designing algorithms to sift through billions of gene sequences, looking for links between certain genetic markers and diseases. The goal is to help us sidestep the diseases we're most likely to contract and to provide each one of us with a cabinet of personalized medicines. Each one should include just the right dosage and the ideal mix of molecules for our bodies. Between these two branches of research, genetic and behavioral, we're being parsed, inside and out. Even the language of the two fields is similar. In a nod to geneticists, Dishman and his team are working to catalog what they call our "behavioral markers." The math is also about the same. Whether they're scrutinizing our strands of DNA or our nightly trips to the bathroom, statisticians are searching for norms, correlations, and anomalies. Dishman prefers his behavioral approach, in part because the market's less crowded. "There are a zillion people following biology," he says, "and too few looking at behavior." His gadgets also have an edge because they can provide basic alerts from day one. The technology indicating whether a person gets out of bed, for example, isn't much more complicated than the sensor that automatically opens a supermarket door. But that nugget of information is valuable. Once we start installing these sensors, and the electronics companies get their foot in the door, the experts can start refining the analysis from simple alerts to sophisticated predictions—perhaps preparing us for the onset of Parkinson's disease or Alzheimer's.

Standing on his magic carpet, Dishman shows me just how deeply the medical Numerati may eventually be able to peer into our lives simply by analyzing our steps as we grab a midnight snack or wash the dishes. He takes a couple of quick paces across the tiles. A video screen behind him displays his weight distribution with a trail of blue and red dots. "Now I'm putting more pressure on this foot," he says, breaking into an exaggerated hobble. "It can tell I'm limping." That might mean that he's had a fall or maybe that his toenails need to be cut. (It sounds silly, but toenail care is something gerontologists keep a close eye on. Untrimmed nails can signal other problems, from immobility to depression to the onset of Alzheimer's. And nail problems can lead to falls, a leading hazard for the elderly and a major focus for Dishman's team. He says that falls in the United States lead to $100 billion in annual medical outlays.) He hops off the tiles and tells me to take a turn. I step on. The tiles are a little squishy, more like cardboard than normal linoleum. (I wonder if they would have absorbed the broken eggs and spilled cocoa that became so common in my parents' kitchen.) The screen behind me displays what Dishman calls my postural sway. It looks like a blue Christmas tree. I see it tilting to the right and quickly make an adjustment. If I were one of the elders currently test-ing the magic carpet, the system would record that Christ-mas tree pattern, establishing it as my "baseline sway." If the pattern changed, Dishman says, it could mean muscle loss or perhaps a side effect from a medicine. "You start capturing this data all the time, and you start to get some really nice trending information," he says. "Basically, every time Mom walks, you want to compare it statistically to every other time she's walked." He lifts his voice a notch. "Wow, Dad's stum-bling more, especially in the morning. Why is that? Is it be-cause of a medication he took overnight that's not trailing off in the morning? Or is it the onset of some cognitive disease?"

The machine won't be able to answer those diagnostic questions, at least not in the near future. It will simply issue alerts when it detects changes in patterns and perhaps urge the user to schedule a medical appointment. It will be up to doctors and nurses to follow up, figuring out why someone is limping or swaying differently at the kitchen sink. But in time, these systems will have enough feedback from thousands of users that they should be able to point people—either doc-tors or patients—to the most probable cause. In this way, they will work like the recommendation engines on Netflix or Am-azon.com, which point people toward books or movies that are popular among customers with similar patterns. (Amazon and Netflix, of course, don't always get it right, and neither will the analysis issuing from the magic carpet. It will only point caregivers toward statistically probable causes.)

Dishman's team has installed magic carpets in the homes of people with neurological disorders or a history of falling. They're starting in the kitchen but would like to extend the tiles into hallways, where they can capture more walking data. They're also trying out two other technologies, a camera that monitors the entire body and a clip-on sensor, about an inch wide, that captures all sorts of data about movement and body tilt. Maybe, says Dishman, the magic carpet can report the data right to the user and act as a fitness coach. He says that the tiles, working with the home computer, will be able to "literally lead you though exercises." And while this is happening, the machine—naturally—is busy collecting even more data. If the user is wearing one of those clip-ons, it might capture heart rate and customize the workout, just like a StairMaster at the gym. (I can't imagine my own mother doing that one in her last years. But perhaps some sprier eighty-somethings would give it a go.)

Thirty-four-year-old Matthai Philipose is one of the quants in Intel's Seattle labs. He and his team pick apart Dishman's data. I call him up and ask how they use statistics to infer our health and behavior from footsteps on a kitchen floor. How far are we from impromptu machine-led calisthenics? He laughs. "These tools weren't even around three or four years ago," he says. To reach the kinds of sophisticated analyses I'm talking about, they'll need to stitch together lots of smaller observations, each with its own range of probabilities. That's what his team is working on now. Start, for example, with a toothbrush. In experiments, the Intel team has wired them with radio tags. These send an alert each time the toothbrush is moved. Can Philipose's group infer that each time it's moved, someone is brushing his teeth? Not necessarily. Someone might move the toothbrush when cleaning the sink. So the statisticians create a chart of toothbrush movement. Let's say they see lots of activity in the morning and at bedtime. Together those two periods might represent 90 percent of toothbrush movement. From that, they can calculate a 90 percent probability that toothbrush movement involves teeth cleaning. (They could factor in time variables, but there's more than enough complexity ahead, as we'll see.) Next they move to the broom and the teakettle, and they ask the same questions. The goal is to build a statistical model for each of us that will infer from a series of observations what we're most likely to be doing.

The toothbrush was easy. For the most part, it sticks to only one job. But consider the kettle. What are the chances that it's being used for tea? Maybe a person uses it to make instant soup (which is more nutritious than tea but dangerously salty for people like my mother). How can the Intel team come up with a probability? One way, of course, would be to survey thousands of homes and ask people what they do with their kettles. That's too much work. Philpose favors a simpler approach.

"Go to Google," he says, "and type 'making tea.' How many Web pages does it locate?" (I do the search. It finds 261,000.) "Then you do another search, but add the word 'kettle,'" he says. (This time it comes up with 29,500.) That gives scientists a rough conditional model. It assumes that of the incidents associated with "making tea," a bit more than one of nine involves kettles. Like so many of the statistical assumptions, it starts as a crude estimate. But it's a way to populate a monstrously big statistical table, one that comes up with the most likely behavior for thousands of scenarios. As more observations pour in from more sensors, the machine itself can adjust and refine the numbers. "Bootstrapping" is Philipose's word for this variety of machine learning. "These kinds of models are good enough that they can start bootstrapping themselves," he says. As they do, they'll come up with better and better guesses about what we're doing every minute of the day.

In these early stages, Philipose says, the Intel team is building statistical models around three groups of observations: morning and bedtime rituals, movement around the house, and nutrition. With these models in hand, they can start adapting the same statistical approaches the Numerati use to look for correlations among shoppers. In this case, do people who use the kettle or the microwave oven at lunchtime have high sodium levels in their blood? How about a person who forgets to brush his teeth more and more, and walks with slower steps through the house? Philipose's numbers won't tell the story by themselves, at least at this stage. But they should point doctors and nurses toward the people most likely to need help.

Now, PICTURE an elderly couple. The husband speaks. The wife says, "What?" He repeats. She still can't make it out. She crosses the room, turns an ear toward him, and what she finally hears is this: "Go get your ears checked." So she does. And it turns out that her hearing isn't the problem. It's that her husband is speaking more softly—perhaps as a result of Parkinson's disease. (And by the time his voice softens to this degree, the disease is advanced.) This is a key area for research because signs of Parkinson's can show up in voice patterns and bodily movements as much as a decade before the disease is usually diagnosed. Early treatment, in exercises and medicines, can delay its development and lessen its impact. Dishman tells me that specialists studying the actor Michael J. Fox in his old TV shows can detect the onset of Parkinson's years before Fox himself knew he had it. His gait grows shorter as the disease creeps up on him. His voice patterns change.

Very few of us have been creating weekly half-hour videos

over the past 20 years that record our changing speech patterns and gestures. But with today's technology, we're in a position to turn the cameras, along with dozens of other sensors, on ourselves. And some of us will start doing just that. Think of all the people who pack their diets with cancer-fighting antioxidants and those who try to stave off heart disease (and risk pneumonia) by jogging relentlessly through the winter slush. Many of us are more than ready to take aggressive action when it comes to lengthening our lives. So it's only natural that at least some of us will train a few sensors on ourselves and send the feeds to Numerati-powered consultants. Anyone offering this type of predictive service in the next few years, I should add, is likely to be a charlatan. For most diseases, the behavioral patterns are not yet established. But once analysts build up a decade or two of data, they'll see the onset of disease early enough, hopefully, to nip it in the bud.

This power to predict is sure to raise a host of social and economic issues. Will those of us who resist using sensors be viewed as reckless, like those today who go for years without getting a medical checkup? Will governments demand a certain level of electronic reporting? Will insurance companies treat unmonitored customers as high risk, denying them coverage or saddling them with the same extortionate rates they levy today on teenage and drunk drivers? These are not issues yet, because the science is at an early stage. But Dishman's team and others around the world are making progress every day.

For now, some of their most useful work is focused on helping people already struggling with diseases such as Parkinson's. In these cases, Dishman says, a stream of data can help doctors refine medications. The status quo in many hospitals is to check patients once a year for 15 minutes to a half-hour and to give them prescriptions based on data gathered in that brief period. This regimen is especially ill suited to Parkinson's, whose symptoms fluctuate wildly even over the course of a single day. "It's a once-in-a-year shot in the dark," Dishman says. "Think about it. You drive there, find a parking space. Your blood pressure is probably through the roof. Then you go into this very unnatural setting where they give you a series of diagnostic tests. And the nasty part of it is that if you're having a particularly bad day, they're going to increase your levodopa, the drug for Parkinson's, which has a bunch of side effects."

In clinical trials, Intel is installing five Parkinson's tests in the homes of people with the disease. Some of these tests are familiar to patients. In one they press two piano-type keys as quickly as possible. Another has them placing tiny red and green pegs into what must be maddeningly small holes. Traditionally, a nurse with a stopwatch measures the time it takes their trembling fingers to finish the job. The Intel version does this electronically and even notes the patterns made as the user drags the peg across the surface of the box in search of the hole. Another device, which looks like a watch, measures the second-by-second shaking of the arm. In this early stage, Intel is simply gathering the data. But the next phase, says Dishman, will be to "close the loop," giving the data to a doctor who can prescribe medicine on a day-to-day basis. He predicts that in time, computers will establish behavioral patterns and make the recommended prescriptions, first in the form of suggestions to the doctor and eventually directly to the patient.

While Dishman's gadgets measure our behavior from outside our bodies, other researchers are busy developing sensors to report on the changing conditions inside. Teams of researchers at the Koch Cancer Institute at MIT are already testing implantable nanosensors in mice. These are built on a scale so infinitesimal that it's hard to imagine. One building block of these sensors, a cone-shaped grouping of molecules called a carbon nanotube, is as small compared to a soccer ball as that ball is to the earth. Tyler Jacks, director of the Koch Institute, says these sensors can detect chemicals in the blood that indicate the growth of a tumor. This technology, potentially, would mean that cancer survivors would not have to wait nervously for their annual checkups to see if the cancer had metastasized (often to an untreatable stage). Instead they'd receive radio alerts immediately, perhaps straight to their cell phones. Doctors could then attack the nascent tumor. Eventually, Jacks envisions placing a host of microscopic sensors into all us, with tools to measure all kinds of conditions and alert us to trouble down the road. For these to work, the Numerati will have to develop statistical norms for hundreds of our biological patterns, from sodium and sugar levels and blood cell counts to the manufacture of all sorts of proteins. These will be our baselines, just like my tilting Christmas tree as I stood on the magic carpet. Developing the most precise models will be especially important as doctors embark on the next step: automatic treatment. "The next generation of embedded medical gizmos," Jacks says, "will know what therapy you need and will deliver it." Included in this micromedical cabinet, he predicts, will be an apothecary of so-called smart bombs, nanoparticles that can be dispatched for precision work, such as attacking cancer cells. It sounds promising, but as you might expect, these new sensors and medications will have to wind their way through years of development, trials, and regulatory approvals before they work their magic inside us. Some animals, though, don't have to wait nearly so long.

· · ·

"THEY SAY it's noninvasive. But I'd feel invaded," Dan Andresen tells me, "if someone stuck a tool kit into my stomach." He points to a big rust-colored steer named Norman, who's wearing what looks like a white plastic Frisbee halfway down his left flank. This is actually a door, a fistula, which opens into the second of his four stomachs. It was surgically attached when Norman embarked on life as a beefy, three-quarter-ton lab rat on this research farm at Kansas State University. Fistulated cows are common in cattle country, where farmers and researchers like to keep an eye on digestive workings. But what passes through Norman's door is most unusual.

Sometime later this spring morning, one of Andresen's grad students will climb into Norman's ring. He'll unscrew the steer's hatch and drop a black plastic packet, about the size of tennis ball, into the sloshing gallons of half-digested alfalfa. Inside that packet is a circuit board rigged with all sorts of technology. It has sensors to measure the temperature and barometric pressure inside the animal. It has a global positioning unit to track Norman's steps in the unlikely event he bolts from this small corral and strolls down the hill through campus to the shady suburban streets of Manhattan, Kansas. Norman's tool kit also includes a wireless transmitter with a small antenna and a memory chip big enough to log the animal's movements and bodily functions. Much of this technology is a work in progress. But one day, when Norman wanders to the trough, the data will fly from his stomach to a wireless receiver, which will send it zipping straight into Andresen's computer.

Dan Andresen, who teaches computer science at Kansas State, grew up on a cattle ranch in eastern Nebraska. About a decade ago, Andresen had lunch with Steve Warren, a software engineering professor who also spent his boyhood around cows. Warren had come to Kansas from the Sandia National

Labs in New Mexico, where he'd been working on health monitors that people could strap onto their arm or around their chest. People, it turned out, weren't the greatest test subjects. The devices were much bulkier back then. And humans often used their advanced brains and opposable thumbs to remove them. What Andresen and Warren needed was a duller and more pliant population. They had their idea before they even paid for their sandwiches.

Together they would build a computer network as vast as the Great Plains. It would stretch from the parched brown pastures of Kansas and the feedlots in Nebraska and Texas to slaughterhouses in Iowa and Minnesota. This network would not only track the health and movements of American cows, but it would also exist on or inside them, perhaps in packets behind their heads or encased in pellets they could swallow. Warren and Andresen planned eventually to put a wireless computer on half a million cows in Kansas — a state where the 7 million cattle outnumber people by nearly three to one. This would produce untold mountains of cow data. Heartbeats, head bobs, munch-munch, a siesta under a shade tree, a glug of water. Run that stream of data 24/7 and multiply it by a half-million, and it would create perhaps the most tedious reality show in the long history of agriculture. But the patterns in that data, analyzed mathematically, could point to all kinds of insights. Who knew what they might find? Perhaps they'd see fluctuations in cows' temperatures before they got sick. Maybe they'd spot an epidemic working its way across the state. The key was this: instead of veterinarians checking up on cows every few months, computers would be reporting on them every single minute.

The two professors drew up a grant proposal and received funding from the National Science Foundation. Following the

terrorist attacks of 2001, interest in the project grew. By tracking every animal from birth to the slaughterhouse, and even
following its subsequent parts and byproducts as they were
transported and sold, authorities could take a big step toward
securing the nation's food supply. In a sense, wiring the cows
would be akin to equipping each animal with a recording machine, like the black boxes airplanes carry. If a diner in a restaurant anywhere in the world bit into an American steak and
fell ill, the trail of information could help authorities trace the
problem not just to a certain region or feedlot, but conceivably to an individual cow. They might see that on a certain
day, while the cow was grazing in a certain Kansas pasture, its
vital signs abruptly changed. That kind of detail could help
solve the mystery.

Isn't it strange, I say to Andresen as we make our way
back from Norman's corral, that we're working to monitor the
health of cows before we get around to people?

"Cows don't care much about privacy," Andresen says.
He's wearing a brown outback hat to keep the sun off his fair
face, which he says burns easily. And he's wearing socks under
his sandals, the way northern Europeans do. "If they care," he
adds, "they don't let on."

Then again, even if cows like Norman knew and cared
about privacy, would it be utterly foolish of them to sacrifice
a dose of it for medical monitoring? Consider yourself. If the
medical industry came up with a system like Andresen's—presumably one that didn't involve clamping a fistula onto your
stomach—would you sign up? This is the kind of question
we're likely to be facing as sensors and computers and wireless
networks grow ever stronger and cheaper. Forget about the
rest of the herd, they'll say. We can create a custom service just
for you—provided you fork over your data.

Already, auto insurers are up to something very similar. In Britain, Norwich Union offers special rates to drivers who agree to place a black box full of recording instruments inside their car. This way, the company can monitor driving behavior and offer further discounts to drivers who keep the speed down and stay clear of high-risk roads and neighborhoods. In other words, the insurers are analyzing not just the drivers' profiles or records, but also their behavior.

This already happens in a rudimentary way in health insurance. Smokers often pay higher premiums, for example. But imagine how much more sophisticated this model could become if we were wired with sensors. New insurance markets could take shape, all of them feeding on the fluctuating signals streaming in from our bodies. In this world, buying health insurance could start to feel like taking out a mortgage. Here's the choice: Lock in a fixed rate, and the company insures you no matter what. But it costs a mint. For lower premiums, you might opt for a floating rate, with premiums that rise and fall with your health risk. Those who play these numbers shrewdly come out on top in medicine, replicating their success in finance. They're data masters. The rest of us underwrite their winnings.

I can just imagine angrily calling the help line when I see on an insurance bill that my rate edged up a few dollars despite the impressive drop in cholesterol. "But your blood alcohol went above the threshold six times," comes a voice from a distant country. He turns a deaf ear to my arguments that red wine is part of the regimen . . .

We're sitting at a table in Andresen's computer lab, which is chock-a-block with components and circuitry and blinking lights. Andresen, his hat hanging behind him from a loop around his neck, is up at the whiteboard, describing a type of cow known as a "dark cutter." These are a scourge. Somewhere along the line—no one knows exactly how or when—dark cutters appear to have experienced some sort of trauma. As a result, their meat is bluish instead of red. It appears to be emptied of its blood. No more T-bones, porterhouse steaks, or top-of-the-line filets mignons from these beasts. The sole option at the slaughterhouse is to grind dark cutters into cheap hamburger. Each one represents lost money.

Now let's imagine that a few years from now, Andresen, Warren, and their team have successfully stitched together their network of cows. A certain number of those animals, inevitably, will turn out to be dark cutters. Researchers will have at their fingertips the lifetime record of every one of these cows, each wag of its head, each snooze in the shade. They'll be able to feed this data through their computers and search for patterns. Do the dark cutters have anything in common? Did they suffer jolts or get too cold on the road trip to the feedlot? Did they sleep less than the others or eat at a different rate? It's all guesswork at this point. But what they discover could eventually lead to adjustments in the way cows are raised or transported. Perhaps certain practices handed down by generations of cowboys will have to be dropped—and replaced by science.

What if the data shows that from their earliest days as calves, dark cutters behave differently? Researchers could turn this knowledge into a predictive tool. This would give them a behavioral profile, etched in math, of a dark cutter. In the same way, each calf could conceivably be scored on the likelihood that it would grow up to produce bad steak. What then? Would ranchers cut their losses and send high-risk calves to be slaughtered right away?

Such questions are at least a few years away—at least

for the cows in Kansas. Building a bovine network is an immensely complicated undertaking. Getting all the sensors to work in sync is a slog, and each one presents its own challenges. The heartbeat, for example, is hard to distinguish from the sounds of fluids and gases going about their noisy business inside the animal. Radio signals struggle to escape thick walls of beef. Batteries wear down and die. Then there are networking conundrums. How do you update software on a thousand head of cattle, or protect them—heaven forbid—from hackers? Still, those are technical issues. Many of them resemble the challenges that engineers mastered in the cell phone industry. If the economic payoff is big enough, they'll work through them.

And when they do, the focus will inevitably move to implanting the devices in us. Governments looking to cut health-care spending will certainly be interested. Electronics companies, as Eric Dishman will attest, view health surveillance as a mouthwatering market. And for the insurance industry, the more information they have about us, the better they'll be able to calculate risk and create a host of new personalized services. Put those groups together, and you have a mighty coalition. Who knows? Given the potential health benefits, many of us may well cheer them on.

LODGED SOMEWHERE on Microsoft's massive computers are a host of e-mails my mother sent to my Hotmail account over the years. She averaged about three a week. A historian or a sociologist could look through them and study the patterns of an American couple early in this century marching purposefully into extreme old age. The e-mails report on Thursday night dinners with the grandchildren and dog walks in the rain. She writes about her activities on the vestry at church and about my father's latest letters to the editor at the *Oregonian,* deploring the treatment of prisoners at Guantánamo Bay. In a couple of the e-mails she writes about taking a cab to a health center in Portland where she and my father participated in a study on aging and cognition, or as she put it matter-of-factly, senility.

Now I find myself at the Oregon Center for Aging and Technology, or Orcatech, a hulking state-of-the-art health center on the banks of the Willamette River. Inside, legions of Portland's elders labor on long rows of treadmills. A café in the spacious sunlit lobby sells expensive lattes. Outside, a ski lift hoists doctors and patients in a gleaming glass pod to a complex of hospitals at the top of a hill. This is where my parents came to lend their brains to science. But in the future, I realize, as I talk to researchers, the elderly can save the cab fare. With the spread of sensors, the cognitive laboratory moves into our homes, where analysts will keep tabs on the working of our brain by tracking the patterns of our daily activities. Nearly everything we do—if studied in meticulous detail—provides a glimpse inside our head. I hear this from researchers constantly. Whether they're discussing the changing pattern of steps on the magic carpet or the adherence to a pharmaceutical regimen, they add, "This also gives us a good cognitive read." It's like a two-for-one sale. Test anything, and you get brain results as a bonus. In this type of analysis, a long string of e-mails, the kind my mother sent to me, would qualify as Exhibit A.

How do analysts correlate irregular footsteps and typos with dementia? The research starts in the laboratory, with the sensors nowhere in site. Micha Pavel, a Czech-born mathematician, explains how he runs seniors through a series of drills over time to test their memory. For each person, he draws up a model of working memory. Like an actuarial chart, it predicts the "survival" of each piece of information. In some, the lines are fairly flat. The memory's holding. In others, it curves sharply. If you look at each forgotten fact as a death, as Pavel does, some of these people are hosting full-blown epidemics. "We try to assess the probability that an item will be lost," he says. "Is it a function of time or intervening events?" In most cases, he says, it's new events that push out memories, as if each person has a limited storage space, what he calls a "memory buffer." Naturally, the people to worry about are those whose buffers are shrinking. Once Pavel has this memory data, the next step is to study the rest of these people's lives — their mouse clicks, word choice, sleeping patterns — and to draw correlations with what's happening inside their head. The work has barely begun. But studying their written words is a natural place to start.

For centuries, people have been scrutinizing letters for insights about their loved ones. When handwriting deteriorated and non sequiturs popped up, they had reason to worry. I certainly fretted as I saw my mother's e-mails grow shorter and less regular. And when I began to see typos in messages from this former legal secretary, I was alarmed. But this was late in the process, only a year or two before she died. Could a rigorous statistical analysis of her typing patterns, sentences, and word choice have pointed to problems years or even a decade earlier? In cases of early detection, doctors can quickly start medicines and therapies to forestall or slow the deterioration. Dishman, meanwhile, is working on a host of technologies to help Alzheimer's sufferers cope. One is a phone prompt. When a friend or family member calls, the person's photo and name pop up on a screen, along with some details, such as the last time that person called.

I walk into the model house in the Orcatech lab. It's strewn with gadgets that they're testing in the homes of scores of Portland's elderly. In one corner is a sensor-wired bed, like the one the little dog jumped on. Lying on the floor is a cane with a boxlike appendage at the bottom. It's designed to measure how much the user leans on it—signaling the possible weakening of that person's legs. Amid these gadgets in the model house are two computers, which I think of as the brain and a separate nerve center. One computer sits hidden in a closet. It picks up all the wireless signals from sensors around the house and relays them back to Orcatech. The other PC sits in plain sight. It's outfitted with a host of games, along with standard word processing and e-mail programs. Every interaction with the computer, every keystroke and click of the mouse, sends back details on cognitive trends. Researchers are still at the early stages now, trying to build a baseline for each user. But within a couple of years, says Pavel, "we're going to be measuring the motor speed, the keyboard interactions, and the complexity of the words they generate."

One model for this analysis is a study of the writings of the prizewinning British novelist Iris Murdoch, who died of Alzheimer's disease in 1999. Murdoch left behind decades of written manuscripts, a treasure trove for cognitive researchers, at University College London. They studied her word choice at different points in her long career and found that in her last novel, *Jackson's Dilemma,* published in 1995, she used a simpler and less varied vocabulary than in earlier works. In fact, they saw that her language followed a curve. It grew more complex from her first novel, *Under the Net,* to one written at the height of her career, *The Sea, the Sea,* before falling off at the end. The scary thing (from my perspective, at least) is that with advanced statistical analysis of different writings, from blog posts to e-mails, researchers (or even employers) may

pick up the downward trend of our cognitive skills long before we even suspect it.

Eric Dishman's team is searching for similar clues in speech
and social interactions. In 300 homes in Oregon, they're piecing together models of people's relationships. How often do
they make phone calls, and to how many people? How often
do friends visit? For each person, they're turning this data into
a score, a so-called "social health index." If people's indexes
fall, it's an indication that something has changed—perhaps
a deepening of dementia. They're also testing a subject's responses to familiar voices on the phone. Usually, people recognize voices of close friends and family members instantly. If
there's a lengthening pause before that recognition, they want
to capture it. "We're looking at milliseconds of difference,"
he says. "That might be an alert that there's some kind of
trouble."

This analysis can get complicated. Picture yourself as
one of Dishman's subjects. You're watching a basketball game
on TV. It's overtime between Dallas and San Antonio. Tony
Parker drives the lane and is fouled. The phone rings and you
reach for it.

"Hello."

"Hi."

That's Nowitzki's sixth foul. He's screaming at the ref.

"HELLO?" comes your sister's voice from the phone.

"Who's this?" you say absently, watching the replay.

"Me, you idiot . . ."

Plenty of things besides the early signs of dementia can
interfere with our thought processes. Music, anger, and sleep-
iness all throw us off our stride, as do cocktails. In time, behavioral scientists will try to incorporate these distractions
into their models. The only way to do it, as you might have
guessed, is by learning even more about us.

Dishman is working on one gadget that takes this to an extreme. It's a wearable device to help people deal with fits of hyperaggressive anger. This involves a heart monitor connected to a souped-up cell phone. A user, Dishman says, starts by filling out an electronic form listing "the people most likely to stress him out." His boss might be one. Next he lists the places that cause stress. "I might be relaxed at the pool," Dishman says, "but more stressed out at Intel." Finally, the user makes his electronic calendar available. By combining all of this with the location data from the cell phone, the key pieces are in place: where the person is, who he's with and what he's up to. So if the heart starts racing, Dishman says, "the system can look at the calendar and say, 'Oh my God, that's a stressful meeting because Kevin's in the room. We've got to get him out of there!'" At that point, the phone rings. The user answers and is prompted by the computer, according to the most dire scenario, to leave the room. Then he's led through a series of questions. "Are you clenching your jaw?" it asks. "Your fists?" It recommends taking some slow breaths or getting a drink of cool water.

For now, the Intel team is trying out this technology on students who don't have hyperaggressive disorder. It gives them cues for managing stress. "I've analyzed your calendar," Dishman says in his machine voice. "And I see you have meetings every half-hour, back to back, for 12 hours. Five of those meetings are with people you've flagged as the most stressful, including your boss and this asshole at work. Do you want to give yourself a break?" Ultimately, he says, this type of data, from our schedules to the people most likely to fill us with rage, should merge with our other health information, including our genetic particulars, the medications we're on, and the feeds streaming from dozens of sensors. He sees each one of us eventually building what he calls a "dashboard to manage our health and wellness." This will be the control panel for our lives. There is almost nothing about us that these dashboards will not be eager to learn. And naturally they'll incorporate all of the trends and medical insights gathered from the dashboards of everyone else.

So here's a question. Let's assume that some form of Dishman's vision takes shape, and each one of us builds a far more detailed repository than we have now of medical and personal data. Who do we share it with? "This comes up in every study we do," he says. "How do you help an elder with Alzheimer's, who's not computer savvy, decide who gets the data and who doesn't? It's a huge design problem. You go into one house, and they say, 'Anyone can have this data.' You go into another, and they say, 'My son can have the data on finances, my daughter can have the data on health, and my other son who I'm pissed at can't have any data about anything.'" Coming up with laws and technologies to help people wisely manage the privacy of medical records is a challenge every bit as daunting as predicting Alzheimer's or short-circuiting a hyperaggressive fit.

As Dishman goes on about privacy, I'm pondering my own medical secrets. Who should I share them with? Then I wonder, more to the point, how eager am I to learn about them myself? As research advances into patterns of disease, behavior, and genetics, we're going to be bombarded with loads of statistical projections about every conceivable malady. Say you learn that you have a 20 percent risk of going blind in old age from macular degeneration. You can lower the odds or delay its onset, you read, by altering your diet, quitting smoking, and taking some pills. Do you change your life to respond to that risk? How about a 7 percent chance of having a stroke in the next ten years? How seriously would you take that? What if your risk in some disease is 8 percent and the national average for your age group is 6 percent? Is that worth paying attention to? Our medical charts, covered with numbers and probabilities, will start to look like the scorecard of a Las Vegas bookie. We'll be awash in the odds of our own demise.

Here's my prediction. As the analysis of our medical data grows, new types of consultancies are going to process these reports for us. Picture a company, SpareMeTheDetails.com. They'll add up all of our reports and give us a life prescription, a combination of medicines, dietary tips, even exercise regimens on the magic carpet, all of them designed to keep various scourges at bay for as long as possible. This process, of course, will involve sophisticated algorithms based on probability, which will create loads of work for the Numerati. But the point is that many of us, in an age of exploding medical data and analysis, may be happy to pay for the privilege of remaining, to one degree or another, in the dark.

CHAPTER 7

Lover

MY WIFE SHOUTS from down in the computer room. She says the online dating site is asking her to describe the man she's hoping to land. "What do I write?" she asks.

"Just describe me," I yell. I'm at the dining room table, filling out the same dating form on my laptop.

I hear her muttering below. She's not happy that I'm dragging her into an experiment that promises to reveal the algorithms of love. I want to see whether an online dating service I'm looking into, Chemistry.com, will sift through the answers and essays we provide, run them through their analysis, and offer each of us to the other as a good match. Later I'll be talking to the Rutgers University anthropologist who devised Chemistry's matchmaking formula, Helen Fisher. She was the one who suggested, in a phone call, that my wife and I try this test. I agreed and strong-armed my wife into joining me. We vowed, naturally, not to respond to any of our fellow daters on the service. We were testing the algorithms, not the people. Granted, there's nothing scientific about studying one couple. For all I know, our marriage is a near-miraculous triumph against the odds.

I embark on this research, I'll admit, with plenty of doubts. An individual's emotions are hard enough to decipher, let alone predict. We've seen in politics how tough it is to figure out what draws people to one party or another. How much more complicated will it be to match two people, each one as complex as a universe? The Numerati's approach works best in areas where sets of consistent data reflect faithfully what they're looking for. Our spending and earning patterns tell them about our risk as debtors. That's easy. Supermarket managers given a list of papaya lovers can bet that a healthy share of them will respond to a discount on mangoes. Piece of cake. But what data best describes us as mates? How can you model someone as a lover?

I remember having dinner a few years back with friends in Pittsburgh. The husband was saying that his wife's brother really blew it when he broke up with his girlfriend. "She was un-be-liev-able," he recalled, shaking his head. He piled on the adjectives. Incredible, amazing. Clearly, there was something about this woman, some piece of data that he viewed as highly desirable, perhaps essential. Would he dare say it in front of his wife, who was sitting right next to him? I finally asked him.

"She had a great . . ." He paused to add emphasis, and I was thinking of all the possibilities. "Personality" would work. If he said "body," or worse, "ass," our friendly dinner was in trouble.

"Job," he said. She was a highly trained nurse who could make good money in any town in the whole country. His brother-in-law, if he had known what was good for him, could have enjoyed guaranteed income wherever he went. (I should note, just to place the comment in context, that this friend's wife had lost her job about a month earlier.)

In theory, finding an online partner for my friend would be a cinch. Just rank the dating prospects by income or credit score, almost as if they were applying for a loan. Then he could start from the top. But I'm betting this nurse he admired had certain other qualities he valued, perhaps ones he wouldn't mention in front of his wife. Maybe he barely recognized them himself.

So how do scientists break down love into pieces that can be fitted into a statistical hierarchy? Love has long resisted such measurement, which is why it has stuck stubbornly to the realm of poets—and frustrated scientists at every step. Shakespeare, no doubt, understood it far better than Newton, and was one of the greatest experts of his day. The challenge today is to find modern experts, whether from anthropology, like Helen Fisher, or psychology, and to team them up with today's Newtons: the Numerati. What they have to do together—and this is the tricky part—is distill what they know about human love and relationships and fit it into a series of algorithms. Some think this is folly. But these online dating sites provide a bonanza of relationship data, more than Shakespeare or even Dr. Kinsey could have imagined. The Numerati are laboring in laboratories of love.

To compare the old, intuitive style to the scientific approach on the website, I send an e-mail to my friend who many years ago, in El Paso, Texas, set me up with my future wife. Which bits of data, I ask him, led him to make this recommendation? I get back a response within minutes. The resulting data, if you could call it that, isn't anything the most gifted of the Numerati could measure or model. He writes that he could communicate well with both of us, that we had a "similar sense of humor," and he felt "an enormous amount of positive energy." Pure fog, from a data-mining perspective.

Small wonder, then, that Helen Fisher and her colleagues call their service Chemistry.com. They're trying to uncover certain data at our essence—perhaps the qualities that Shakespeare compared to "a summer day" or "a worm 'n the bud"—and package it in matchmaking algorithms.

I slog through scores of questions on the Chemistry site, while working to quell a mounting insurrection from downstairs. Clearly, many are designed to gauge whether I'm outgoing, adventuresome, cautious, a stickler for details. Some, though, are harder to read. I'm shown a photo of a man and a woman having a drink on a terrace. Brother and sister? Lovers? Husband and wife? I guess lovers, but it's hard to know what that answer says about me. Then I'm asked to compare the length of my second and fourth fingers. That too is a mystery. The form asks about relationships that have thrived. I describe my wife. Finally, I get to the essay where I'm asked to define myself. This scientific system, I'm assuming, will analyze the word selection and the length and syntax of my sentences, and then use them to pigeonhole me as one type of lover or another. My choice of words will even reveal my inner secrets, I've been told. But for this to work, I figure the system has to see the free-flowing, unedited me. So I relax and, for their computer's benefit, type copiously about myself. I write about a year I spent long ago in South America as a near hermit. I write about this book, and about going to coffee shops and putting on noise-canceling headphones and writing. I go on and on with this stream of consciousness. Then I hit the send button. A few minutes later, I'm appalled to see this blather appear verbatim on the profile, right next to the blank space where my photo would appear. This essay was no quiet tête-à-tête with a computer! It was ad copy for myself. What does that have to do with data-crunching Numerati?

They're only collaborators in this effort. Much of it, at least at this stage, involves self-promotion. It's the same process we go through when we apply for the class of 2014 at Harvard or try to land a six-figure consulting job at McKinsey and Co. We put forth the most flattering image—much as we have throughout history. And we look for the Numerati to find us the right match. But what if we're both lying?

That's the risk we take. Now that we're advertising ourselves, many of us may choose to litter our profiles with little white lies and deal with the consequences later. What counts, as any advertiser will attest, are results. Look at my possibilities. I have a 1999 picture of myself that might give a good bounce to my profile on Chemistry.com. If I Photoshopped away that small fin-de-siècle zit on my nose, lightened the bags under my eyes, and added $50,000 to my reported income, who knows what the youthful, wealthier (and not entirely honest) me could accomplish on this site?

Is this science? That depends. Consider the data I provide to Chemistry.com. I start with demographic details, and I hand them over knowingly. I have a pretty good idea how they'll be interpreted. Some may flatter me. Others I part with in the resigned spirit of full disclosure, or a believable facsimile. So my prospective dates learn my age, address, profession, education, religion, even my income if I choose to provide it (which I don't). The questionnaire asks if I have children living at home. (It politely refrains from asking if my children and I are living there with my wife.) It then asks about the women I'm willing to consider. Do I prefer to exclude certain religions or body types? (Exclude no one, I say.) Am I interested in teetotalers, high school dropouts, or women taller than me? (Yes, I want them all.) These types of questions date from the infancy of computer dating, in the 1960s. They don't demand sophisticated analysis or modeling. They only ask the computer to carry out simple matching chores, to put us into piles. These are the details marketers and politicians have feasted on for decades. Many still swear by them. And these preferences are accompanied by what many consider to be the most important piece of data in the entire process: the photograph. (I withhold mine. This may raise suspicions among my dating prospects, but I'm interested in how the Numerati interpret my data. They're not analyzing the photos, though perhaps they will in years to come.)

Next comes a stream of involuntary data that was unavailable to the punch-card pioneers of computer dating. This is our behavior on the Chemistry.com site. Analysts at the company, like their colleagues throughout the e-commerce universe, record our every click. They can measure which types of potential dates appear to interest us the most. Then they can showcase more of the same genre to us (and people like us). They can break us down, category by category, and look for trends. This analysis is nearly identical to that done by Tacoda's Dave Morgan and other online advertisers. They don't pretend to know us in any depth. Our thumping hearts and quivering gills remain a mystery to them. They simply count our clicks, study our behavior, and then put us in buckets and market to us.

Finally, we venture into the newest realm of data: the survey responses, which, when interpreted by scientists like Helen Fisher, create our love profile. This is where most of us lose control of the process. It's hard to figure out what kind of self-portrait we're creating because many of the questions seem mysterious. But who wants to fine-tune at this juncture, anyway? If we're paying money to find the ideal date, most of us (at least in theory) want the service to understand us as well as it possibly can. So we scrutinize the length of our fingers, puzzle over that picture of the smiling couple on the terrace, and fudge only in cases where the question appears to probe something we'd rather not admit to. (If the questionnaire lays traps for child molesters or pornographers, I haven't spotted them.) The idea in this psychological section is to advance far beyond demographics and behavior, to burrow deep within us, and untangle our heartstrings. This is to understand us at our most basic level—as beings engaged in mating rituals we share with other animals, from shad to kangaroos. I have no doubt that this test will help Helen Fisher and her team understand at least some of the cravings and neuroses operating inside me. But do these profiles really line us up with the right person? Or do they just give us something to talk about on our first date? That's what I'm out to learn.

So I ask Helen Fisher when I catch up with her by phone. Fisher is both an anthropologist and a neuroscientist. Matchmaking, she says, "is the most important game we play. From a Darwinian perspective, if you have four children and I have none, your genes win." And she thinks that the standard data used in matchmaking sites, the shared hobbies and interests, are nearly worthless for finding a spouse. "You can have the same ethnic background, the same socioeconomic standing, the same general level of intelligence," she says. "Your looks, religion, politics, and goals can all be aligned. You can walk into a room full of people like that, and you don't fall in love with them. I can't tell you," she adds, "how many relationships I walked out of where the person was perfect on paper." She's confident that her method will decode the human lover just the way other efforts I've been telling her about model us as shoppers, voters, and workers. "We're going to get to the bottom of this, just like IBM and Yahoo," she says. "The human animal does have patterns."

Fisher says that in the late 1990s she began looking into the biology of personality, the genes, neurotransmitters, and specifically, the hormones. She did this in part by studying brain scans of "romantically obsessed" people. Her theory is that four different hormones—estrogen, testosterone, dopamine, and serotonin—mold our personalities and that we look for people who complement us, who provide what we're missing. Her questionnaire is designed to divide us into four different types, each one with a dominant hormone. Some of the questions focus on the moods and personalities she associates with each hormone. Others, such as the question about the length of our fingers, zero in on the chemical itself. Research shows, she says, that those with an index finger shorter than the ring finger have often been exposed to more testosterone while in the womb, while those with longer index fingers will have more estrogen.

Fisher outlines the different hormones and personalities for me. Those with lots of dopamine, she says, are likely to be "Explorers," optimistic risk takers. Serotonin breeds "Builders," who tend to be calm and organized and work well in groups. Those brimming with testosterone she calls "Directors." Two thirds of them are men. They're analytical, logical, and often musical. (They sound suspiciously like Numerati to me.) In the fourth group, their brains coursing with estrogen, are the Negotiators. They're verbal and intuitive, and have good people skills. You'd think they'd be built for relationships. But sometimes, Fisher says, "they're so pliable that they turn into placaters. You don't know who they are."

People leave personality footprints everywhere, Fisher tells me, even in the sentences they write. She gives me common words used by each group. Explorers use words like *excite, spirit, dream, fire,* and *search,* while more community-minded Negotiators talk about *links, bonds, love, team,* and *participate.*

Builders are more liable to discuss *law, honor, limits,* and *honesty.* And that Numerati-infested bucket of Directors? Their words focus largely on the physical world, where *aim, measure, strong, hard,* and *slash* have currency. Not surprisingly, they also talk a lot about "thinking."

My wife and I, we learn later in the afternoon, are both Explorer-Negotiators. (Each person gets a dominant and a secondary label.) This sounds promising enough. "You tend to be focused and resourceful, and you are able to juggle a lot of projects at the same time," I read. As a result, we're both "sometimes a whirlwind of activity." But a pairing of Explorers, Fisher warns me, can be risky. "Explorers fly off in different directions the minute they get bored," she says. "They get into relationships fast, wonder how they got there, and then try to weasel their way out."

Okay. Maybe each of us really needs a no-nonsense Builder to keep our finances in order, map our vacations, and make sure the cats have their latest battery of rabies shots. Perhaps that would make sense. But is that what our hearts secretly ache for? As Fisher starts out, her evidence is mostly anecdotal. She describes a classic match. Picture a hard-driving man, a fabulously successful business executive. He bangs heads, slashes the payroll, drives would-be challengers into oblivion. This guy gets things done. He's a Director. And chances are, Fisher says, he has a smooth-talking, problem-solving wife, who quietly patches together all the friendships he shatters. She's a Negotiator. Those two types, Fisher says, "are very symbiotic. They will gravitate toward each other."

Clearly, the service doesn't sense the same gravitational pull between my wife and me. When I log on to the website, I find a list of five women who have the right levels of serotonin and estrogen for people like me. My wife isn't one of them. There's

an insurance manager from West Orange—a Negotiator-
Explorer—who says "we all have to laugh every day, especially at ourselves." A Negotiator-Builder, from Rochelle Park,
works in information security and likes ballroom dancing.
These and three others are the machine's choices. Many other
subscribers, however, have access to my profile. And regardless of the chemistry, they're free to express interest. Whether
they're Builder-Directors from Tarrytown or fellow Explorer-
Negotiators from Toms River, I learn that each one is a "great
match." It gets to the point, as I click through my prospective matches, where the word *great* starts to sound pretty ordinary.

What gives? The automatic system, in all modesty, recognizes its limits and bows to the human brain. As the science
stands now, it can make introductory suggestions. But far be
it for a machine, at least at this point, to overrule the vastly
more sophisticated human and nix a potential Romeo. "Great
match," it says.

Even if it doesn't dare second-guess our judgments, the
Chemistry computer should be able to suggest smarter pair-
ings as it sees which combinations work. Fisher tells me she
has data from 1.6 million people who took the test. She can
see which types of people are most likely to pursue which others. Statistics indicate, as she predicted, that Negotiators gravitate toward Directors, and vice versa. Explorers are attracted
to Negotiators. No-nonsense Builders are often drawn to Explorers, who help them "lighten up," Fisher says. But just as
often, Builders opt for a less combustible combination and
seek out their own kind. With these insights, she can refine
the recommendations—and perhaps lead the system to help
me find my wife.

Of course, the personality groupings are only one smid-

gen of data in our dating profiles. I talk to the analysts at
Match.com, the parent company of Chemistry.com. Each
of us, they say, instinctively seeks out a match from our own
cultural and educational level. We can find this compatibility
simply by reading the other person's essay. As our level of education rises, we use larger words and longer sentences. Daters
naturally tend to select people at their own level. Dating services can accelerate this process by presenting us first to people
with similar vocabularies. We also focus on other similarities,
no matter how silly. The Match.com team has found that if
they can show people that a potential match has three things
in common, interest soars. "We can send them e-mail saying
that there's someone else in their town who also likes dogs and
whose favorite color is red," says Jim Talbott, manager of web
analytics. He speculates that people look at these as "a little
piece of coincidence, a little piece of fate." Regardless of the
cause, it's easy to hunt down these similarities, use them as
marketing tools, and then measure which combinations appeal most to each type of customer. The dynamic, once again,
is nearly identical to targeting advertising. This proves to be a
much comfier niche for the Numerati than the tangles of human attraction.

Just one difference. In this case, the advertised goods are
us. We want to find, and we want to be found. And increasingly, we're going to have to figure out how to use these statistical profiles of ourselves to create the spark. As our mating rituals migrate from bars and study halls to electronic
networks, honing our algorithms may become as important
as the smiles, scents, and sideways glances that Shakespeare
knew so well.

. . .

HAVE YOU EVER noticed the little button on the Google search engine called "I'm Feeling Lucky"? Type in a query and click that button, and it delivers only a single Web page, the one deemed most likely to satisfy your search. "I'm Feeling Lucky" cuts to the chase. Yet it receives almost no attention and accounts for far less than 1 percent of all searches, according to Google. Why is this? For starters, we don't trust the machine to understand our instructions and deliver the right Web page 100 percent of the time. And what if there's another Web page that's just a little bit better? The fact is, we like choices. We like to browse through the possibilities. Imagine if Chemistry.com, like the matchmakers for European royalty in centuries past, lined us up with just one potential mate. We'd feel cheated. And just like the eternally frustrated Henry VIII, we'd naturally wonder if there wasn't someone with an extra pinch of this or dose of that. Even if science—whether a search engine or an online dating site—had the smarts to give us exactly what we're looking for, we wouldn't really be sold until we saw the other possibilities. (Some of us like to keep testing the possibilities long after we've made our choice.) The key for these services is to give us a selection of good choices.

And the key for us? Our success in the networked world, whether we're hoping to land a date or a job, hinges not just on our ability to find, but also to pop up on the first page of other people's search results. Throughout history, we've developed all sorts of ways to be found. We wear perfume, jewelry, tattoos, platform shoes, all of them delivering a message of who we are. We write up résumés, we build big and complex social networks. We crack jokes. Some of us pay to appear in reference books, like *Who's Who*. As the Numerati assert their ways, we will be located less by the sights and sounds we produce, or even our friendships, and more by math-based programs churning through our data. The trick, increasingly, will be to help the machines find us and to use machines to locate others.

For companies, this burning need to be found has spawned an entire consulting industry. It's called search engine optimization, or SEO for short. Let's say you have a bed-and-breakfast in Tucson. But when potential customers type "Tucson Bed Breakfast" into a search engine, your site doesn't appear until the fifth page. That spells disaster. Potential customers never find you. So you go to consultants and you pay them to engineer your website so that it shows up near the top of the list. (The Internet is teeming with companies offering this service.)

To optimize your page, they have to understand the search algorithm. How does it define a high-ranked page? Is it a plethora of links to other pages? Loads of traffic? The prominence of certain words? Good consultants test thousands of combinations and figure out what the algorithm is looking for. Then they tweak their Web pages to satisfy it. Engineers at the search engines, meanwhile, tinker with their algorithms to put these manipulators in their place and to keep the most relevant sites at the top of their lists. They get feedback on every click. It's an eternal battle, not only between the search engines and the consultants, but also among the consultants. Some of them game Google better than others.

This is something we humans have been doing since our early days as bipeds. Gaming systems is our specialty. We figure out how things work. Then we calculate the necessary steps so that they work for us. This is true whether we're putting together an investment portfolio or angling to win "employee of the year." Each one involves figuring out the ideal recipe—or algorithm—for the intended result. The dynamic hasn't changed. But these days, more of the tricks involve automatic systems. Time was, for example, that before applying for jobs we would meticulously lay out our résumés with just the right fonts and on the best watermarked paper, to attract the attention of a human resources manager. Now, according to *BusinessWeek*, 94 percent of U.S. corporations ask for electronic résumés. They use software to sift through them, picking out a selection of "finalists" for human managers to consider. What does the software look for? That's what we have to figure out. Some pick out certain words—MBA, Harvard, Excel, fluent Mandarin. Others look for more sophisticated combinations. Plenty of consultants are on call to sell us inside tips. The point is that when we want to be found, whether we're looking for money or love, we must make ourselves intelligible to machines. We need good page rank. We must fit ourselves to algorithms.

REACH INTO YOUR pocket or purse, and pull out that blinking cell phone. Have a good look at it. In the past decade, we've come to take these miraculous pocket computers for granted. But they're bristling with radio signals, sensors, computing power, and storage. In lots of ways, they're similar to that tennis-ball-sized packet of hardware bobbing around in the stomach of Norman, the fistulated cow. Now imagine we wanted to emulate Norman. What if we used our phones, like Norman's in-stomach computer, to record our movements and our interactions, and then enlisted some Numerati to create a mathematical profile of each of us? Could we then perhaps find other people with similar patterns? Could those people become our friends and allies—or lovers?

That's what Nathan Eagle thinks. A few years ago, Eagle,

then a Ph.D. student at MIT's Media Lab, tried an experiment. He distributed cell phones to 100 grad students, one quarter of them at MIT's Sloan School of Management, the rest at the Media Lab. These phones, he informed the participants, were equipped with software to record their movements and interactions. Over the course of an entire school year, this data would show the researchers not only where the students went and how they communicated with each other, but also who they circulated with, and even who they were spending the night with. This was a privacy invasion huge enough to agitate a congressional oversight committee. But all of Eagle's subjects signed lengthy consent forms.

By the time I catch up to Eagle, he's living on the coast of Kenya, working on an education project. Our Skype connection fails, so we chat online. He tells me that he's watching turtles swimming in to lay their eggs on the beach below, where they're hard-pressed to protect them from local poachers. "You can get six shillings for these eggs in town," he writes. I nudge him toward faraway Cambridge, and he tells me about his experiment. Over the school year, he says, it became easy to see that the two groups of subjects—the business school students and the engineers—moved in different patterns. He could predict with greater than 90 percent certainty which type of student each one was. What's more, he could look at different types of relationships and figure out which people were friends and which were mere acquaintances. If they met at the water cooler on a Thursday afternoon, that was one kind of relationship. If they were together at a bar on Saturday night in downtown Boston, they were much chummier.

Eagle began to build models of the individuals. He started with basic patterns of cell phone use: whether the grad students were at home or at work, if they kept their phone on or off. Each of these variables was called an "eigenbehavior." (The prefix *eigen* is a multiplier of an established trend or direction.) It was easy to calculate the mean for each of these behaviors. The users, when charted, fell into clusters. That's how Eagle distinguished business students from engineers. Even within those clusters, each individual had a unique combination of behaviors. Some slept past noon on Saturdays. Some kept their phones off Sunday mornings (church?). When Eagle mapped them in colorful charts, each individual's life looked as orderly as the geometric forms of a Navajo rug. They were so regular, in fact, that he could predict with fair accuracy what each person would do next. He could predict where each one would go, who they would call, what time they would turn in for the night, and whether they would bother turning off the cell phone when they did.

All kinds of organizations are hungry for such data. Mass transit companies want to predict the movements of commuters. Local advertisers, naturally, would love to hit a phone user with an ad for a bar or a restaurant right as they're getting ready to carouse. And I don't have to tell you how useful the Homeland Security Department would find this tracking data.

But Eagle has the idea that we can put this data to use for ourselves. He wants to go into the friendship business. Imagine, he says, that we can switch on our telephones to a "promiscuous" setting. This means we're open to chance encounters. Our phone works as a beacon, sending out our profile in radio waves to those around us. In the early days, this profile will be like the early days of computer dating. It will include a list of our interests. Swedish movies, say, or bicycle touring, French food. And if those whose paths we cross share these of these interests, our profile will pop up on their phones, and we presumably won't mind at all when one of them touches our elbow and says, "I had a coq au vin to die for at this little bistro . . ." In the workplace, a similar system could alert us to colleagues in the cafeteria who have mastered the Linux operating system or are knee-deep in the genetics of drosophila flies.

But take this a step further. Our movements with a cell phone can paint an in-depth profile for each of us, each one endlessly more detailed than those forms my wife and I filled out for Chemistry.com. If we give them permission to examine us the way Dan Andresen and his team study their cows, they can scrutinize our movements and social networks. They can map the DNA of our behavior. Why would we give anyone a green light to do this? Imagine that they could use this data to find other people whose profiles match our own. Would they become our next friends? The love of our life? That ocean of mobile data may well be the next frontier for Helen Fisher and the other matchmakers of the world.

Already, companies are amassing loads of this data. Consider that cell phone I asked you to pull out of your pocket a few minutes ago. Your phone carrier can detect that it's sitting right there, unused. And the company has more than enough information to draw powerful conclusions about you and to make predictions (most of them centered on the chances that you'll jump to another carrier). It's a potential gold mine of personal data. Phone companies have all they need to track our movements and our social networks. They could analyze the photos many of us send and the words of our text messages. As we surf the Net and begin to use the phone for e-commerce, they learn even more. If they wanted to (and were ready to face a wave of lawsuits from privacy advocates), they could build entire businesses on this rich data. Or perhaps at some point they could package and sell our own data right back to us. That's Nathan Eagle's idea. His scheme, which is only in its infancy, is about empowering us — using our data to make ourselves happier, richer, and surrounded by more friends — or perhaps just to know ourselves better.

THE EARLY RESULTS from Chemistry.com cast my marriage in an ominous light. Neither of us pops up on the other's Chemistry.com radar. On an otherwise bright Sunday morning, I see the service is lining up my wife with a Negotiator-Director in Rosedale, New York. Calling himself "Working Class Hero," he writes, "I stop to smell the roses to find the beauty that is in all of us regardless of the soil the roses were planted in." Perhaps those words resonate for my wife, a horticulturist. The chemistry between the two, according to the computer, also bodes a little too well: "With the spontaneity and creativity of the Explorer and the flexibility and imagination of the Negotiator, you're both in store for some great adventures and hearty laughs together." I'm willing at this point to recognize Working Class Hero as a worthy rival. But Rosedale? It's 40 miles away, on the far side of New York City (and within whiffing distance of JFK Airport). That's a slog. Google tells me that the drive could take "up to 1 hour and 50 minutes in traffic." Is this logistical nightmare, for all its potential, really preferable to a connection with an Explorer-Negotiator named Stephen who lives in the very same town (and, as luck would have it, in the same house)?

I recheck my profile to see if there's some detail that's keeping us apart. And that's when I see it. When filling out the form, I had carelessly limited my search to women younger than my wife. Silly me. I was blocking the only connection that mattered to me and practically throwing my wife into the arms of more open-minded rivals. It was data under my own control that betrayed me.

I promptly raise my age limit. Within hours, my wife checks the Chemistry.com page, and there I am. Finally, it's the photo-less Stephen from Montclair, a fellow Explorer-Negotiator wooing her with the same goofy essay about wearing noise-canceling headphones in cafés. Naturally, the service declares this match "great." And our Explorer-to-Explorer connection brims with promise. "You'd both be happy jetting off to Paris or Nepal at a moment's notice. And your life in the bedroom is likely to be exciting too." At this point, my wife designates Stephen as "sizzling"—and the rest of my rivals as "fizzling." The algorithms have completed their work. The suitors have been banished to Bernardsville, Rosedale, and beyond.

In truth, we set up something of a farce, and it's easy to laugh about it. But whether our hormones and the length of our fingers had anything to do with it, the service actually made good on its most important challenge. It allowed us to find each other. What we make of the rip-roaring possibilities ahead is up to us.

CONCLUSION

ONE SUMMER long ago, Terry Therneau is telling me, he went to the shores of a lake in northern Minnesota to work as a counselor in a summer camp. Therneau is a leader in quantitative biology at the Mayo Clinic, and this detour from medical data into the forests of his youth takes me by surprise. "One of my goals that summer," he says, "was to learn the name of every tree in the woods." It seemed like a simple enough proposition. From what he could see, there were a few dozen to learn. He continues: "As I learned the trees, I began to see more and more of them. Pretty soon, my estimate of the woody plants went up tenfold." Complexity appeared to grow with his knowledge. Now he sees the same phenomenon as he studies the human body. Millions of proteins, all of them interrelated, swarm through our cells like "clouds of gnats." The more he learns, the more he sees. When Therneau went back home that summer, he still didn't know all the trees in the northern woods of Minnesota.

The Numerati too are grappling with towering complexity. They're looking for patterns in data that describe something almost hopelessly complex: human life and behavior. The audacity of their mission is almost maddening. They're

going to figure out who we're likely to vote for, who we want
to work with, perhaps even who we're best suited to love, all
from the statistical patterns of data? It's the height of presumption, and it leads to humbling disappointments. Like the trees
growing in the forests of Minnesota, we confound those who
try to categorize us, and we do it most of the time without
even trying. Life is complex.

And yet, bit by bit, the Numerati make progress. No,
they don't truly know us, and they never will. But in each domain, they understand and predict our behavior a bit better
today than they did last week. They learn from their mistakes.
They haul in more data. They continue to experiment. This
is a scientific process, and from the laboratories of advertising
to counterterrorism, each of us is laid out as a specimen. In
some, we're rendered in fine detail. In others, it's bare bones.
But there's no turning back from the trend. In the age we're
entering, our lives will be described, studied, and predicted,
every day more, through this statistical analysis.

This will lead to all sorts of frustrations. We'll be confronted, from time to time, with conclusions that are ques-
tionable or even flat-out wrong—but delivered with the certainty of science. Sometimes, this will reduce our options.
Already, insurance companies equipped with statistical analyses of costs and survival rates are imposing their mandates
on doctors, who were once freer to trust their gut. That trend
will only grow. And as the numbers proliferate and researchers learn more about our DNA, hospitals, insurers, and government agencies will be instructed by automatic systems to
discriminate. Like discerning gatekeepers at exclusive clubs,
they'll wave in certain people before putting a hand up and
saying, "Not you."

How do we fight back? With numbers. For this, we have

to understand the methods that produce these analyses, and we must master some of them ourselves. In the past, for example, a worker might make the case for a raise with a pithy paragraph in a year-end review. (That's been my approach.) Now, increasingly, those of us who can quantify annual achievements on an Excel spreadsheet have the edge. In the most dire cases, we'll hire lawyers who have mastered the tools of the Numerati and who can debunk faulty and self-serving conclusions drawn from statistical curves and correlations. The battle, whether it's at work or in the courtroom, revolves around the analysis of data.

As this world takes shape, we'll have to figure out how much of ourselves to hide. Decades ago, I'm told, my sister-in-law grappled with this question. She was stepping out of the shower in the bathroom of her all-women's dorm, and she heard the call "Men on the floor!" At many schools, this would have been a non-event, but she was in a highly conservative religious college. She was naked. She had only a small towel to cover herself, and there were men prowling the hallways. She could hear them. She waited, but they didn't go away. So she began to think about which part of her body to cover with the towel. It barely fit across her bottom or her top. It certainly didn't cover both. She had to make a choice. Finally, she had an inspired idea. She threw the towel over her head—and then scampered naked to her room. Given the options, it was more important for her to cloak her identity than her body.

In this new world, all of us are going to face situations in which our most intimate data is exposed, at least to somebody. And we may be interested, or at least willing, to share some of this data. HIV-infected patients, for example, might want to participate in a study and reveal lots of information about their symptoms and their spirits, maybe even their habits, but under one vital condition: that they remain nameless. The personal data can be shared but not the identity.

So we're going to have to reevaluate our ideas about privacy and secrets. We all have different types of secrets. Some things we tell no one. Others we share with family and a friend or two. Many are secrets in name only; we blab about them all the time. But until recently, our secrets were scattered. The doctor guarded some of them, the banker others. The high school teacher, the dressmaker, the neighbors, the office mates, they all had their allotment. Some existed only in their memories, with certain details escaping, from time to time, into rising and falling streams of gossip. Many were scrawled on receipts or prescriptions, police forms or warnings from school. Most of them, if we played it right, didn't mingle much. Unless a detective was on the case, the bits of information didn't find each other. Now they can and will.

This can be scary. No doubt it will tempt a few of us to turn away from the data-spewing world altogether. Some will tiptoe around the Internet, if they venture there at all. They'll pay with cash, avoiding the trail of credit cards. They might even wait in long lines to throw coins into the toll machines instead of coasting through the automatic readers (which can track many of our movements and even calculate our average speed).

But with a bit of knowledge, we can turn these tools to our advantage. You may not have noticed, but as we make our way in these pages from the snooping workplace to the laboratories of love, we gradually evolve from data serfs into data masters. In the beginning, employers are using the tools to analyze and optimize us as workers. In many of their calculations, we might as well be machines. Advertisers and political operatives gather up our data to plunk us into buckets. But they're doing it to provide us with more ads and promotions targeted to our tastes and values, to give us more of what we want. That's a step toward power. Once we're in Intel's home-health labs, hitching sensors to our bodies and wiring our kitchen floors with magic carpets, the balance shifts. We're appealing to the science of the Numerati to protect us from falls and alert us before strokes and heart attacks. And by the time we're prowling for love on Chemistry.com, we've come full circle. We're paying for algorithmic profiles of ourselves and lining up mathematical correlations with potential mates. The point is, these statistical tools are going to be quietly assuming more and more power in our lives. We might as well learn how to grab the controls and use them for own interests.

Where to start? In these early years, it's hard. It involves reading the tiniest print on privacy disclosures at e-commerce sites and on the back of credit-card applications. But as we learn more about the value of our data, and our vulnerabilities, we'll no doubt clamor for services to help us manage it. That should attract businesses to serve a growing market. One nonprofit organization founded in 2005, AttentionTrust, is leading the way. It provides Web surfers with the tools to amass their own data and to sell it, if they choose, to advertisers. In essence, AttentionTrust is urging people to harvest their own clicks and words—and to stop giving them away to companies like Tacoda, Umbria, and countless others. AttentionTrust hasn't yet spread much beyond a circle of the Net savvy. And so far, the markets for selling our own data are embryonic. But that could well change as the broader public learns more about how the Numerati are adding us up.

* * *

As I TYPE on a Sunday afternoon, I put on my noise-canceling headphones and jack up a Mahler symphony—all to block out a loud tutoring session upstairs. My 15-year-old is plowing through algebra. This leads me to wonder what he'll need to learn for a life in which he'll be measured in a thousands ways, analyzed bit by bit, and then reassembled and optimized by statistical wizards. Does he need to tackle advanced calculus? Should he delve into operations research, learn to manipulate eigenvectors and hidden Markov models? Do he and millions of others need to become Numerati themselves?

In a word, no. Let's start by taking on three of the enduring myths that misinform this discussion and have done so for centuries, or even longer:

1. The world is divided between word people and number people.

This becomes true only because we let ourselves believe it. Mathematicians and computer scientists, in fact, speak words. Many of the ones I've met along my journey were speaking to me in their second or third language. Quite a few were eloquent. And those of us who cloister ourselves on the word side of the divide, who turn the page in a book every time a formula brimming with Greek letters and parentheses pops up (I'm guilty here), we too have minds full of numbers. We're constantly adding and dividing and carrying out processes whose mathematical names sound alien to many of us. Consider this example. The baby woke up crying at 11 and again at 1, and then at 2:30. Does this mean—we lie in bed doing a quiet regression analysis—that the next cry will come at 3:30?

The key difference between the Numerati and the rest of us lies in the toolbox they carry. It contains sets of mathematical formulas and drawers full of algorithms that mankind has been building for thousands of years. Using this know-how, they attempt to put complex reality into numbers so that theories can be tested and refined. They analyze whether new buildings will stand or bombs will explode, and they handle those traditional tasks on their own, with minimal input from those of us who aren't handy with such tools (and cringe when confronted with them).

But the new challenges are different. The Numerati must now predict how we humans will respond to car advertisements or a wage hike. The models they build will fall flat if they fail to understand human behavior—if they plug in the wrong data. Figuring out how to boil us down to numbers requires not only the right tools but also the real-world context. That means they must work in teams that draw from different disciplines and include people with all kinds of expertise. There's plenty of work for anthropologists, linguists, even historians. If there was ever a divide between so-called numbers people and word people, the challenges ahead demolish it.

2. The Numerati are in control. They'll have their
   way with us.

Wrong. Even the greatest and most powerful of the Numerati only master certain domains. Everywhere else, they'll be just like the rest of us: objects of study. Larry Page, for example, is a cofounder of Google and a titan in the world of the Numerati. His scientists are building machines to crunch hundreds of billions of our search queries and clicks, and to sell us, in neatly organized buckets, to advertisers. But when Josh Gotbaum's political program pours through consumer data and classifies millions of California voters, it plunks Larry Page into a bucket of Still Waters or Right Clicks. Whether they're patients with a genetic predisposition for blindness or supermarket shoppers with a sky-high tendency to throw a candy bar in the cart, the Numerati are sitting in the databases with the rest of us.

This is a wonderful thing because the people in the best position to exploit our privacy are also gaining an intimate understanding of how their own privacy can be trampled. They understand it better than anyone. This is the dynamic that turned Jeff Jonas, the Las Vegas data maven, into a privacy advocate.

3. Those who master the numbers will make all
   the money.

They'll make money, no doubt. But not all of it, not by any stretch. Think back to the dawn of the automobile age. In workshops in Detroit and Stuttgart, engineers were turning out new machines poised to change the course of history. But plenty of people who didn't know the difference between a piston and an alternator stood to make fortunes from cars. They just had to understand the trends and plan their businesses accordingly. Some of them built suburban subdivisions, malls, and fast-food restaurants where folks could eat in their cars. Some bought land where the highways would be coming through, and others sold oil tankers too huge to squeeze through the Panama Canal. Entertainment empires grew around Formula One racing and NASCAR. The motor economy was open to those who saw where things were headed.

That is as true today as it ever was. To make the case, follow me into one more business, Inform Technologies. Its founder, a former banker named Neal Goldman, is hard at work building his second fortune. He's no algorithm wizard. But he has the imagination to see what the Numerati can build, and he has shown an uncanny ability to find good ones.

In the 1990s, when he was in his twenties, Goldman was an up-and-comer for Lehman Brothers in New York. He worked 120 hours a week, doing cross-border mergers and acquisitions, some of them worth billions of dollars. "It was incredibly intense," he says. He would regularly pull all-nighters, preparing for early-morning presentations to management. So he'd be pecking away at the computer, getting numbers from Bloomberg, facts from analyst reports, details from annual reports. He was pulling together data, and it took time. "I'd spend a few hours organizing it, putting it into an Excel spreadsheet. Then at around 3 A.M.," he recalls, "I'd start thinking." How absurd it was, he realized, that a highly paid professional was spending most of the night hunting down data and plugging it into a spreadsheet. "Out of a 12-hour process, I was spending one hour thinking," he says.

Goldman saw these all-night headaches at Lehman Brothers as a looming business opportunity. So he quit in 1998 and started a company. His plan was to build a tool to organize and structure all the diverse bits of information he spent those nights hunting. All of the connections had to be a click or two away. Someone studying an investment in a steel mill, for example, should be able to find not only the financial records and stock performance of steel companies—that was easy—but also the key players in the industry, their background, and articles written about them. He should also be able to track the people in the companies, where they'd worked before and gone to school, the connections they shared with board members. The service he envisioned would stitch together the entire web of the world economy, from raw materials to personal relations. For this, he would have to place an immense jumble of information into the same symbolic universe. Goldman was no mathematician, but he knew that if all these pieces of data were going to swim together in the same pool, they would need to be represented in a common language. He needed a geek.

Goldman advertised on a website, and one day a 16-year-old high school student named Joe Einhorn knocked on his door. "He was so shy, he couldn't look you in the eye," Goldman says. For a test, he gave the boy "some undoable tasks." Einhorn reappeared a couple of days later. "He'd stayed up for 48 hours and coded the crap out of the thing." Goldman had found a greenhorn ready to join the ranks of the Numerati. Joe Einhorn was his first employee. Later, Joe's brother Jack would sign on. Since the age of 13, he'd been doing cancer research at a New York University program at Veterans Administration, looking for statistical patterns in the expression of a gene involved in the development of prostate cancer.

Eventually the team grew. New partners, investors, and tech specialists climbed aboard. The tool Goldman envisioned, Capital IQ, started to take shape. And it worked. Much of the financial universe was represented within it, in a complex matrix of vectors. All of the data circulated in the same orb, and it was organized by relationships. Want to find Yale grads sitting on corporate boards? Click. How about former Enron executives in the energy business? Click click. Goldman found customers for it, and in 2004, he and his partners sold the company to the Standard & Poor's unit of McGraw-Hill for $225 million.

By the time I catch up with Goldman, he's onto his next start-up. His new venture, Inform Technologies, is a Numerati-fueled precision weapon targeted at many of the people

I work with: editors. At its root, Inform is much like Capital IQ. It ventures into the tangled, multilingual universe of written news, and it proposes to match readers with the webs of far-flung stories that will interest them. In its early stage, Inform sets out to organize the entire world of news so that every article is linked to every other piece of news that relates to it. A single profile of the Venezuelan strongman Hugo Chávez leads readers along lots of related trails, one about the oil industry, another about revolutions in Latin America, a third about Chávez's friends and allies in Moscow and Tehran, and one about his rocky relations with Washington. In Inform's scheme, each piece of news is a thread connecting to an immense and constantly mutating tapestry depicting the world today. It's ambitious. But that's just the beginning. In time, the idea is to follow the readers' clicks and queries and turn them—or us—into statistical profiles. Each one of us will then get a customized stream of news. To create this service, the Inform team—led by the Einhorn brothers—must bring the world's news onto a single mathematical platform. As Jack Einhorn describes it to me, Inform's universe of news exists as a sphere of infinite dimensions, with the stories shooting through it as vectors. Each one intersects with the names and themes that they include. Related stories travel in the same clusters in this imaginary space. They intersect. This is similar to the vector-ridden galaxy we encountered in Carnegie Mellon's "next friend" analysis. But this time, instead of scouring your social networks for a French-speaking lawyer, it may be tracking down just the article you want on changes in French law.

When I think about where I fit in this algorithmic economy, I need look no further than Goldman's start-up, which carries the Numerati straight into the heart of journalism. The editor he's building has far more range than the human versions I've known. In the world in which I've built my career, most of the reporters covering school board meetings, tornados, floods, and wars are young. The idea is that when they grow older and more settled, they'll be promoted to a higher-paying editing job. In theory, their long experience will help them pick and shape stories that serve and interest their readers. Call it judgment or gut feeling, it's what sets these editors apart. But as I climb up to Inform's sixth-floor offices, overlooking 57th Street on Manhattan's East Side, I'm looking at an operation built to automate editing. If machines handle the editing, what should human editors do? Study math?

Let's take a closer look at Inform. I walk into the office, and there are 30 workers in four low rows, all of them hunched over computers. None of them look at me. After a minute, Joe Einhorn greets me. He's in his midtwenties and wears a baseball cap. He leads me into a conference room and asks if I want something to drink. It takes me a minute to realize that I've just met the chief science officer. By this point, his baseball cap is bobbing far down one of the aisles. I plug in my laptop and wait for Neal Goldman and Joe's brother Jack.

Goldman, who's in his late thirties, wears dark brown hair parted near the middle. The silver zipper on his black turtleneck reaches up to his Adam's apple. Like Dave Morgan and Tacoda and Howard Kaushansky at Umbria, he harnesses the power of the Numerati without mastering their science. No doubt, he knows a whole lot more math than a liberal arts major like me. He has an MBA and worked in global finance. He understands statistical analysis. But he's no whiz at set theory, algebraic geometry, or computer science. He harnesses his imagination, which is crucial, and he delegates. "I understand it conceptually," he says. "I can take a problem, and I can start to parse out what types of data or scoring it would involve. Then I communicate it to people like Jack."

Goldman's skills are a lot more relevant, from a career-planning perspective, than the Einhorns'. Those brothers have special ability. Most of us don't have it and never will. Societies must engage and nurture such talents. These are people, after all, who will be building the next Googles, perhaps vanquishing fearsome forms of cancer and exposing hidden networks of terrorists. For every society, locating these people at a young age, whether they're living in barrios or working in rice paddies, is an imperative. But that's a policy issue. For the great majority of us, becoming a math or computer science prodigy is not a career option.

So what other jobs are there? I look down those long rows of programmers at Inform and ask Jack Einhorn what kind of skills these people have. Some of them are wizards in their own right, he says. One, his childhood friend Ray, is building autonomous robots to go sniffing out news. Another, a Chinese Ph.D. named Kai, specializes is experimenting with algorithms borrowed from facial recognition technology to pick out similarities in news articles from all over the world. Other workers, both in New York and India, are working on more mundane tasks. They're building small applications, much like a toolmaker in an auto plant.

Some of these people work—as they say in the software business—with their heads up. They talk to their colleagues on the floor, with designers, maybe even with users or the sales team. They collaborate. As they do, their value rises. Pull them out of the operation and a whole series of connections goes dead. Others are known in the profession as head-down. They're on their own. In the world we're entering, head-down workers who are not highly skilled are vulnerable. Since they are not knitted into the greater fabric of the project, they can be more easily replaced, like stand-alone machines, by cheaper head-down workers offshore. After all, numbers and computer codes zip offshore far faster than auto plants do. Math is no safe haven. Only those who like it, and are good at it, should pursue it.

And the rest of us? We should grasp the basics of math and statistics—certainly better than most of us do today—but still follow what we love. The world doesn't need millions of mediocre mathematicians, and there's plenty of opportunity for specialists in other fields. Even in the heart of the math economy, at IBM Research, geometers and engineers work on teams with linguists and anthropologists and cognitive psychologists. They detail the behavior of humans to those who are trying to build mathematical models of it. All of these ventures, from Samer Takriti's gang at IBM to the secretive researchers laboring behind the barricades at the National Security Agency, feed from the knowledge and smarts of diverse groups. The key to finding a place on such world-class teams is not necessarily to become a math whiz but to become a whiz at something. And that something should be in an area that sparks the most enthusiasm and creativity within each of us. Somewhere on those teams, of course, whether it's in advertising, publishing, counterterrorism, or medical research, there will be at least a few Numerati. They'll be the ones distilling this knowledge into numbers and symbols and feeding them to their powerful tools.

IT'S A SUNNY morning in Palo Alto. I'm having breakfast with a venture capitalist and then a session with Google in the afternoon. The cell phone rings. It's my old college roommate, a Ph.D. in computer science, who probably forgets every year or two more math than I ever learned. I'm on my very first days on this odyssey, and I have the excitement of someone who's just stepped into a new world. I tell him what I'm up to and explain, in a sentence or two, how the mathematicians are going to dip into the sea of data to form models of all of us. "This is the mathematical modeling of humanity," I say. The connection starts to fill with static, but before it goes dead, I hear him say, "I'm going to call you back!"

A few minutes later, I'm driving north on U.S. 280, looking for the Sandhill Road exit, when the phone rings again. "I'm really worried about your story," he says. I tell him that I'm heading to interviews, too busy with driving to talk. He tells me to pull over. After I do, he explains that once he too dreamed of modeling the world but has since concluded that math, while powerful, is flawed.

"Why?"

"Ever hear of garbage in, garbage out?" His point is that mathematicians model misunderstandings of the world, often using the data at hand instead of chasing down the hidden facts. He tells the story of a drunk looking for his keys on a dark night under a streetlight. He's looking for them under that lamp not necessarily because he dropped them there but because it's the only place with light.

Later that afternoon, I'm sitting at an outdoor patio with Craig Silverstein, Google's chief technologist. He was the number-one employee at Google. The founders, Larry Page and Sergey Brin, hired him because neither one of them, for all their brilliant ideas, knew much about search engines. It's sunny and the wind is blowing the pages of my notebook, and I tell Silverstein the story about the drunk looking for his keys.

He smiles. He's heard it many times before. He recalls a science fair in junior high, where his project featured lots of good data he'd come up with. "I wanted this data to be significant," he says. "Finally I found a test that matched it." The judges, he adds, weren't fooled.

Spending all this time among the Numerati, I've found myself wondering what jobs the rest of the world will handle in an economy dominated by calculations. Now it occurs to me: it's up to us to help them find the keys. The mathematicians and computer scientists create magic but only if their formulas contain real, meaningful information from the physical world we inhabit. That's the way it's always been, and even as they mine truckloads of data, it's a team effort. "In the end," Silverstein says, "you're just counting things."

What's new, of course, is that many of these "things" the Numerati are busy counting are people. They're adding us up every which way, and they have all of humanity to model. The rise of this counting elite will convulse entire industries. It's already happening. At the same time, I suspect it will lead many of us to give more thought to who we are. As we encounter mathematical models built to predict our behavior and divine our deepest desires, it's only human for each of us to ask, "Did they get it right? Is that really me?"

ACKNOWLEDGMENTS

> <

NOTES

> <

SOURCES AND
FURTHER READING

> <

INDEX

ACKNOWLEDGMENTS

I'd like to thank my colleagues at *BusinessWeek,* who gave me so much help and support through this process. It was editor in chief Steve Adler who suggested doing a cover story on math. I thank him for giving me the story and later granting me more than a year's leave of absence to report and write this book. Neil Gross was my thoughtful and patient editor for the magazine story and a sounding board as I worked on the book. Thanks also to Peter Elstrom, Frank Comes, and John Byrne. My agent, Jim Levine, jumped on this book idea with great enthusiasm and helped tremendously with the proposal. Elizabeth Fisher worked tirelessly on the foreign sales.

I couldn't have dreamed for a better editor at Houghton Mifflin than Amanda Cook. She helped shape the book conceptually. It started early in the process when she told me that practically the entire book was sitting in Chapter 4 of my proposal. She was right. During the writing stage, she sent a steady stream of marked-up manuscripts from Boston to Montclair while assuring me on the phone that things really were moving ahead. Thanks also to Susanna Brougham for helping to shape the final text. I'm grateful for the wonderful work of Bridget Marmion, Lori Glazer, Patrice Taddonio, Sanj Kharbanda, and Elizabeth Lee in promoting the book.

I appreciate all the generous help given to me throughout the process by my sources for this book, both those cited in these pages and the many more who are not. A special thanks to Anne Watzman at Carnegie Mellon University, who introduced me to the World Wide Web and coaxed me in my Pittsburgh days to shift my focus from steel to technology. Thanks too to my neighbor and favorite mathematician, Alfredo Bequillard.

I'd like to direct a salute to my parents, Mary Jane and Walter, who were both thrilled when the project began and were fully involved in spirit to the very last sentence. My sisters, Judy, Sally, and Carol, were loving and supportive throughout. Thanks as well to my sons, Aidan, Jack, and Henry, and to my wife, Jalaire. She endured the computer-dating ordeal, among other things. And she put up with the clicking of the laptop (which drives her crazy to this day) and my reclining form on the blue couch for more months than either of us would like to count.

NOTES

INTRODUCTION

2   *Gorges on data. Data* is a plural form of the singular noun *datum*. But in many fields, *data* is treated as a singular noun, just as the singular word *sand* stands for lots of individual bits of silica. That's how I use *data* in this book.

5   *In a single month, Yahoo alone.* See "To Aim Ads, Web Is Keeping Closer Eye on You," *New York Times*, March 10, 2008.

6   *Crack mathematicians.* A word about the genesis of this book. I was pitching a cover story at an editorial meeting at *BusinessWeek*. The story, I said, would focus on the risks to the technology economy of the United States. Many regions of the world had better Internet connections and superior wireless networks, and they were graduating more scientists and engineers. What's more, the U.S. economy, which had long attracted some of the brainiest foreigners, was blocking more of them from coming to our country post-9/11. And those same foreigners had plenty of lucrative opportunities at home. The editors found this pitch too familiar. Wasn't there a fresher way to attack this story? Neil Gross, a senior editor, mentioned that mathematics was at the heart of most of the key technologies. That idea turned into a cover story. Math was fresh. Who wrote cover stories about mathematics?

I began interviewing mathematicians at MIT, the Courant School at New York University, and Bell Labs. And it quickly became clear to me that writing a story about math was akin to writing one about words. The subject was too vast. So I focused on data, and as I did so, the story swerved from pure math to computer science. Math is a big part of what the Numerati do, though, and so we kept it in the article title: "Math

Will Rock Your World." See http://www.businessweek.com/magazine/content/06_04/b3968001.htm.

7   *Later, notes Tobias Dantzig.* See Dantzig, *Number: The Language of Science,* p. 7.

8   *A mathematician named Bill Fair.* Researchers at Fair Isaac are hoping to use their data-modeling expertise for applications far beyond finance. One potential market is medicine. People who neglect to take their prescribed medications wind up more often at the emergency room with more serious (and expensive) problems. Forgotten or ignored prescriptions, according to Fair Isaac, cost insurance companies an estimated $15.2 billion annually in the United States. So researchers at the company are developing a system to assign to each of us a score denoting the risk that we won't take our pills.

Which details from our life predict that we'll become medical slackers? It might have to do with age, years of schooling, or whether we live alone. There may be statistical correlations among ethnic groups. Fair Isaac's researchers are poring over the data now. But if in the future they figure out how to predict this risk, we might eventually carry a numerical score for so-called prescription noncompliance. Those of us with high numbers might get a call every day or two from the doctor's office, reminding us to take our drugs. Maybe they'll even send someone to knock on our door. Sure it's pricy, but from the insurance companies' perspective, it's cheaper than three weeks in the intensive care unit.

Fair Isaac is also considering creating numerical scores for all sorts of human qualities, including honesty, generosity, and reliability. Employers, of course, would be interested in some of these numbers. And if the philanthropy industry had access to our generosity scores, they could target more efficient fundraising drives. So far, these are still just ideas.

13 *Composed of numbers, vectors, and algorithms.* The word *algorithm* comes from the name of a ninth-century Persian scientist, Al-Khwarizmi. But algorithms were commonplace long before him. Think of an algorithm as a set of instructions, or a recipe. Brenda Dietrich, the chief of math research at IBM, finds them even on the back of shampoo bottles. "Wash, rinse, repeat. That's an algorithm," she says. Algorithms power search engines and marketing campaigns. They schedule the entire major league baseball season. They mete out the hops and barley in a vat of Heineken and the corn syrup and caramel coloring in a tankard of Coke (that algorithm's a carefully guarded secret).

The algorithm didn't rise to its current stardom until the invention of the computer—a machine that demands logical and well-ordered instructions (and is utterly useless without them). With its arrival, an entire branch of applied math and engineering started creating ever more algorithms. These were instructions for counting things, sorting them, making calculations and comparisons, and in short, accomplishing computing tasks. Naturally, many algorithms are packed with statistical analysis. A search engine algorithm, for example, counts how many pages link to each Web page, how often they're each viewed, and how many times and how prominently the key words appear on each page. It builds a hierarchy on a load of calculations. But the instructions, the foundation of the algorithm, are not based on what we usually think of as mathematics. The keys are clarity and logic, within a rigid set of rules. Lawyers, I'm often told, are good at writing algorithms.

One nugget from IBM Research: Many algorithms developed in past decades were viewed as theoretical. But with the dramatic advance in computing power, some of them now can be tried out on computers. They're migrating from theory to practice. This is leading researchers to comb through their archives for hidden algorithmic gems.

15 *Agreed to sell Tacoda to AOL.* Tacoda is not the only behavioral advertising company to get scooped up by an Internet giant. In September 2007, Yahoo paid $300 million for BlueLithium, a start-up very much like Tacoda. And the previous May, Microsoft spent $6 billion for aQuantive, an advertising technology company with a behavioral division, DrivePM.

1. WORKER

31 *He attempted to prove that monogamy.* Dantzig, "Discrete-Variable Extremum Problems," *Operations Research,* vol. 5, no. 2, April 1957.

35 *Mountains of facts about each employee.* "International Isn't Just IBM's First Name," *BusinessWeek,* Jan. 28, 2008. This article reports that IBM has also developed a search engine called Small Blue to locate suitable employees. "The software scans employees' blogs, e-mail, instant messages, and reports, then draws conclusions about each participant's skills and expertise. When other employees search by topic on Small Blue, the program scans its findings to get a list of experts."

40 *It's getting late in Takriti's office.* Samer Takriti left his job at IBM in August 2007 to take a position on the math team at Goldman Sachs. We met for lunch later that fall near his office, not far from the ferry terminal at the southern tip of Manhattan. He said that he was ready for a

change and had briefly considered other job offers, both at rival banks and at Google. He said he was excited to be working in finance and actively involved in business. It has a faster pace than research. Work on modeling the 50,000 consultants proceeds apace, say officials at IBM.

2. SHOPPER

42 *They can study our patterns of consumption.* One path to understanding humans, oddly enough, starts with so-called horse races, statistical tests to compare our behavior with that of others. They're a standard of Internet marketing and a hand-me-down from the direct mail industry. In fact, every time we receive a pile of junk mail, we sort through a herd of test horses. Fair Isaac is a leader in helping companies analyze the results. Larry Rosenberger, Fair Isaac's vice president for research, described the process to me one autumn afternoon at his offices in San Rafael, California.

Rosenberger walked to the whiteboard to show me how credit card companies run these races. He drew a tall tube. "That's a customer," he said. He started drawing lines across it, as if creating little segments on a worm. "You might know his age, gender, income, you might have details on his behavior, what he bought and when. Each of these fields is something about the customer." He drew another segmented worm, this one representing all the variations in interest rates, penalties, and frequent-flier miles that the credit card company offers. (He called it "the offering vector.") Companies test each type of offer, some more generous, some less, with each demographic and then study the results. Eventually, the company can figure out the most profitable combination of incentives and rates—and even the wording and design of the pitch—for each group. Actually, I shouldn't say "most" profitable because these companies always keep testing it against others. Horse races never stop. As we continue to use credit cards, they build up more detailed models of our buying behavior—and send us more horses. They produce more and more data, which can be matched against our growing record—what we buy, where we go, how deep we dive into debt. Some companies have taken this to extremes. Capital One, a leader in microtargeting, has developed more than 100,000 different profiles for credit card offers.
Retail took a half-century detour. For a more detailed description of the post–World War II mass economy, see *The New Marketing Paradigm,* by Don E. Schultz, Stanley I. Tannenbaum, and Robert F. Lauterborn.

44 *Ghani made a splash in 2002.* "Mining the Web to Add Semantic Details

to Data Mining," *Springer Lecture Notes in Artificial Intelligence,* vol. 3209, 2004.

55   *Think of buckets as genes.* The language of genetics pervades the science of the Numerati. I did a search through my notes and found the word *genome* mentioned 139 times. The architectural term *blueprint,* which in the figurative sense is a synonym, popped up only 13 times. In one example among many, Martin Remy, the chief technologist at the San Francisco search start-up Sphere, says that his team develops "document genomes," a combination of features "that lets us find other genetic matchings for documents."

58   *One researcher at Microsoft.* For more on Heckerman, read "Using Spam Blockers to Target HIV, Too," *BusinessWeek,* Oct. 1, 2007.

3. VOTER

68   *Joined with two coauthors to detail this triumph.* See *Applebee's America,* by Matthew J. Dowd, Ron Fournier, and Douglas B. Sosnick, Simon & Schuster, 2006.

69   *Each sliver of the electorate.* Even political data mavens disagree about the value of microtargeting. My reporting took me to the Washington offices of Hal Malchow, a consultant who began data mining for voters back in the 1990s. This made him a mossback among the political Numerati. He said that despite all the excitement about consumer data, the most useful variables remained those that my father might have recognized as he worked to turn out voters for Richard Nixon in 1960. "These six things matter the most," Malchow told me:

   1. Ethnicity. (Whites, blacks, Jews, and Catholics have different voting patterns.)
   2. Gender. (In recent presidential races, a majority of men have voted Republican.)
   3. Marital status. (Democrats do best among single women in a landslide.)
   4. Church attendance. (The pious are more conservative.)
   5. Gun ownership. (Conservatives, with a libertarian streak, tend to own guns.)
   6. Geography. (The higher the population density, the more liberal the voters.)

Microtargeters do not challenge the significance of this list but insist that they can pry loose atypical individuals within these groups.

Malchow, by contrast, argued that many efforts outside of these core areas amounted to marketing hype.

Another note from Malchow: Although African Americans represent a core Democratic constituency, the party lacks reliable lists of black voters. "The myth is that we have African Americans," he said. "We don't." Unlike Hispanics, they don't have distinctive surnames. This leads list builders to search for first names they associate with African Americans, such as Latisha and Jamal. In the process, of course, they miss millions of Roberts, Janes, Toms, and Alices.

75 *As Robert O'Harrow Jr. writes.* See *No Place to Hide,* by Robert O'Harrow Jr., Free Press, 2005.

Mike Henry, Clinton's deputy campaign manager, left the race on February 13, 2008, following Clinton's losses to Senator Barack Obama in Virginia, Maryland, and the District of Columbia.

4. BLOGGER

106 *Companies and governments alike are poring over.* This is happening in countless ways. Consider Michael Cavaretta. He runs a math shop at Ford. He and his team are attempting to mine the company's vast collection of warranty claims. The big challenge is to reduce millions of documents, some of them handwritten, into math. But first the machines must figure out the writers. What do thousands of mechanics and customer service reps around the world mean when they write phrases like "squeak and squeal," "shimmy and shake"? Are those pairs of words synonyms? Should they go into the same bucket? Do the meanings of these words vary by region? Cavaretta told me that one mechanic wrote that a car was "squealing like the pig Bubba stuck." How does a computer make sense of that? Cavaretta's team extracts all the knowledge it can from this vast collection before clustering the data and using statistical analysis to find patterns of problems in the cars.

119 *A blog about deodorants in Iraq.* Stephen Baker, Blogspotting.net, "Captive Advertising Audience at 30,000 Feet," http://www.businessweek. com/the_thread/blogspotting/archives/2007/01/captive_adverti.html.

5. TERRORIST

124 USA Today *reported.* "NSA Has Massive Database of Americans' Phone Calls," *USA Today,* May 11, 2006.

125 *There's a lack of historical record.* This is a problem for NASA as well.

David Danks, a philosophy professor at Carnegie Mellon University, told me that NASA processes data from 40,000 different sensors on the space shuttles, much of it coming in numerous times per second. This provides sufficient data to create detailed simulations of launches. And yet during the first quarter-century of shuttle flights, there have been only two disasters. "We have a sample size of two," he said. This makes it difficult to pick out patterns of data that point to problems.

126 *Unexpected earth-shaking events.* Nassim Nicholas Taleb, *The Black Swan: The Impact of the Highly Improbable,* Random House, 2007.
Jerry Friedman, a statistics professor at Stanford. See *The Mathematical Sciences' Role in Homeland Security: Proceedings of a Workshop,* National Academies Press, 2004.

131 *Jeff Jonas, like many others.* Jonas writes at length about security and privacy challenges surrounding data on his blog, http://www.jeffjonas.typepad .com/.

143 *As many as 300 cameras.* "Watching You Watching Me," *New Statesman,* Oct. 2, 2006.
The Chinese government announced plans. "China Enacting a High-Tech Plan to Track People," *New York Times,* Aug. 12, 2007.

6. PATIENT

160 *"There are a zillion people following biology."* For that same reason, I decided not to focus the medicine chapter on what the Numerati are doing in the vast field of genetics. But I did research the subject. One of my ideas was to figure out the genetic odds that, like my father, I would develop glaucoma and macular degeneration and eventually go blind late in my life. This question led me to the University of Iowa, where a personable doctor named Edwin Stone has built a world-class eye research operation, including the Carver Family Center for Macular Degeneration. I learned there about an experiment to decode the entire genome of a rat's eye, which is similar—despite its beady appearance—to our own. The job for the Numerati studying the rat gene is not to find single genes that create blindness. Those are rare. Instead, the challenge is to untangle tens of millions of relationships among the genes and to map the paths of power and influence within the eye. The secrets to blindness are not found in the structure of the genome but instead in the behavior of its components. It's like a society.

The analysis, of course, is statistical. And as I learned about it, I began to see that it's very much like the work that goes on at Tacoda. Just as Dave Morgan was searching for the behavioral patterns of romantic-movie lovers, genetic researchers have to parse the behavior of the influential genes. What activates them? Are there stimuli coming from other genes or proteins? Which ones? In both domains, advertising and genetics, the process involves sifting through massive sets of data, looking for patterns, weighing statistics, and using probability to distinguish between a cause and a coincidence. From the point of view of the Numerati, the microscopic forces within our bodies behave more like communities, or even markets, than like components of a machine.

I'm sorry to report that I learned nothing about the chances that I would go blind, much less that genetic fixes were at hand. Instead, Dr. Stone prepared me for a gradual approach to battling inherited diseases: "A couple of years ago," he told me, "we identified this gene called the fibulin 5. It's responsible for 1.5 percent of age-related macular degeneration." He made a tiny space between two fingers. "It's this dinky little thing, right?" But the discovery, he said, gives researchers a look at the mechanism that causes macular degeneration. "This allows us then to do experiments that say, now why is that? Why is a tiny change in this gene causing people to get these accumulations under their retinas? . . . If we could understand that pathway," he said, "maybe there are things we could do when somebody is 35 years old to knock that pathway down a little bit. Then instead of the average age of someone losing vision from macular degeneration being 67 or 71 or something, maybe it could be 87 or 91. We'd like it to be never. But from a population point of view, every three or four years that you could move that curve make a dramatic difference in the amount of blindness out there."

171   *Provided you fork over your data.* Hospitals that figure out how to make intelligent use of patient data are bound to rise to the top. This, as I learned on a visit to the Mayo Clinic in Rochester, Minnesota, has long been the case. I met with the clinic's data expert, Dr. Christopher Chute, who told me about a crucial breakthrough. In the first years of the clinic, more than a century ago, he told me, the Mayo brothers ran their operation much like other big clinics. Say a patient came in with a sore shoulder. He was sent to the orthopedic specialist. But it turned out to be a heart problem! So off he went to the coronary specialist. He took some medicine there and broke out in hives. Next stop, dermatologist. Each of these three doctors had a separate record of the patient. Often they had to track down their colleagues to piece together the twists and turns of their patients' cases.

Enter the Mayo brothers' partner, Henry Plummer. In 1907, he and his assistant, Mabel Root, devised a new system. Upon signing in, each patient received a dossier, to be carried from doctor to doctor. This way, each doctor could study the medical history of their patients from the first day they arrived at the clinic. When the patients checked out, their dossiers went into a big file. Plummer and Root put color codings on the dossiers for each type of disease and treatment. The result, said Chute, using language that sounds more Google than Mayo, "They had a paper database that was structured and searchable!" Through the years, they indexed the dossiers with ever finer detail. This enabled them to engage in what our generation would call analytics. They could look at every case of colon cancer or tonsillitis, and analyze which treatments were most effective and cost-efficient. "This was continuous quality improvement," Chute said, referring to the industrial process Japanese automakers made famous decades later. They turned the practice of medicine from a boutique business of independent consultants into a modern business. "This place exploded out of the corn fields." The challenge now, of course, is to come up with a similar breakthrough for medical data in the twenty-first century.

172 *In Britain, Norwich Union offers.* "Norwich Union Buys Tracking Equipment for Pay-as-You-Go Motor Insurance," *Insurance Business Review,* Oct. 6, 2005.

7. LOVER

195 *94 percent of U.S. corporations.* "The Art of the Online Résumé," *BusinessWeek,* May 7, 2007.

196 *Software to record their movements and interactions.* "Gadgets That Know Your Next Move," *Technology Review,* Nov. 1, 2006.

CONCLUSION

215 *"Garbage in, garbage out."* Not everyone agrees with the familiar thesis of garbage in, garbage out. Early in my research, I was talking about it with William Pulleyblank, IBM's vice president in charge of business optimization and a former director of the company's Deep Computing Institute. "Garbage in, garbage out isn't correct anymore," he said. "You haven't got time to clean up your data. The real challenge is how you make something of value from 'garbage.'" In other words, in a fast-moving business world, quick and dirty conclusions have a fighting chance to work. Slow and sure, by contrast, is often an oxymoron, because data may be out of date by the time it's cleaned and vetted.

SOURCES AND FURTHER READING

Ayres, Ian. *Supercrunchers: Why Thinking-by-Numbers Is the New Way to Be Smart*. Bantam, 2007

Barabasi, Albert-Laszlo. *Linked*. Plume/The Penguin Group, 2003

Bardi, Jason. *Socrates: The Calculus Wars*. Thunder's Mount Press, 2006

Briggs, Rex, and Greg Stuart. *What Sticks*. Kaplan Publishing, 2006

Brin, David. *The Transparent Society*. Basic Books, 1998

Courant, Richard, and Herbert Robbins (revised by Ian Stewart). *What Is Mathematics?* Oxford University Press, 1996 (originally published in 1941)

Dantzig, Tobias. *Number: The Language of Science*. Fourth edition. The Free Press, 1967

Gleick, James. *Isaac Newton*. Vintage Books, 2003

Hamm, Steve. *Bangalore Tiger*. McGraw-Hill, 2007

Henshaw, John M. *Does Measurement Measure Up?* The Johns Hopkins University Press, 2006

Morville, Peter. *Ambient Findability: What We Find Changes Who We Become*. O'Reilly Media, 2005

O'Harrow, Robert Jr. *No Place to Hide*. Free Press, 2005

Schultz, Don E., Stanley I. Tannenbaum, and Robert F. Lauterborn. *The New Marketing Paradigm: Integrated Marketing Communications*. NTC Business Books, 1994

Sosnik, Douglas B., Matthew J. Dowd, and Ron Fournier. *Applebee's America*. Simon & Schuster, 2006

Stakutis, Chris, and John Webster. *Inescapable Data: Harnessing the Power of Convergence.* IBM Press, 2005

Watts, Duncan J. *Six Degrees: The Science of a Connected Age.* Norton, 2003

Whitehead, Alfred North. *Introduction to Mathematics.* Barnes & Noble Books, 2005 (originally published in 1911)

INDEX